teach ®
yourself

basic mathematics

teach
yourself

basic mathematics

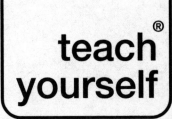

basic mathematics
alan graham

For UK order enquiries: please contact Bookpoint Ltd, 130 Milton Park, Abingdon, Oxon OX14 4SB. Telephone: +44 (0) 1235 827720. Fax: +44 (0) 1235 400454. Lines are open 09.00–18.00, Monday to Saturday, with a 24-hour message answering service. Details about our titles and how to order are available at www.teachyourself.co.uk

For USA order enquiries: please contact McGraw-Hill Customer Services, PO Box 545, Blacklick, OH 43004-0545, USA. Telephone: 1-800-722-4726. Fax: 1-614-755-5645.

For Canada order enquiries: please contact McGraw-Hill Ryerson Ltd, 300 Water St, Whitby, Ontario L1N 9B6, Canada. Telephone: 905 430 5000. Fax: 905 430 5020.

Long renowned as the authoritative source for self-guided learning – with more than 40 million copies sold worldwide – the **teach yourself** series includes over 300 titles in the fields of languages, crafts, hobbies, business, computing and education.

British Library Cataloguing in Publication Data: a catalogue record for this title is available from the British Library.

Library of Congress Catalog Card Number: on file.

First published in UK 1995 by Hodder Education, 338 Euston Road, London, NW1 3BH.

First published in US 1997 by Contemporary Books, a Division of the McGraw-Hill Companies, 1 Prudential Plaza, 130 East Randolph Street, Chicago, IL. 60601 USA.

This edition published 2003.

The **teach yourself** name is a registered trade mark of Hodder Headline.

Typeset by Transet Ltd, Leamington Spa, England
Printed in Great Britain for Hodder Education, a division of Hodder Headline, 338 Euston Road, London NW1 3BH, by Cox & Wyman Ltd, Reading, Berkshire.

Hodder Headline's policy is to use papers that are natural, renewable and recyclable products and made from wood grown in sustainable forests. The logging and manufacturing processes are expected to conform to the environmental regulations of the country of origin.

Impression number   10 9 8 7 6 5 4 3
Year   2009 2008 2007 2006 2005

contents

acknowledgements

Many thanks to Wendy Austen, Carrie Graham, James Griffin and Sally Kenny for their help in the preparation of this book.

part one

one

understanding the

basics

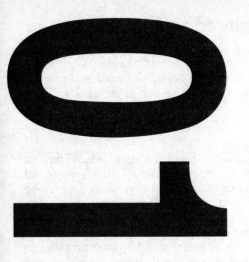

01

reasons to be cheerful about mathematics

Is this book really for me?

I wonder what made you decide to buy this book! No doubt each individual has his or her own special reason. As author, I clearly can't meet everyone's exact needs, so I tried to write this book with two particular types of reader in mind. I will call them Marti and Mel.

Marti

Marti was neither particularly good nor particularly bad at mathematics, but was able to get by. She could usually work out the sums set by her teacher at school, but never really understood why they worked. She had a vague sense that there was, potentially, an exciting mental world of mathematical ideas to be explored but somehow it never happened for her. She now has two children aged 10 and 6. She is aware that her own attitudes to mathematics are being passed on all the time to her children and she wants to be able to encourage and help them more effectively. The crunch for Marti came when her daughter asked her about the difference between odd and even numbers. Marti knew which numbers were odd and which were even, but she couldn't really explain why. She might have bought this book to understand some of the basic 'whys' in mathematics.

Mel

Mel never got on with mathematics in school. He lost confidence with it at an early stage and constantly had the feeling of 'if only the teacher and the other pupils knew how little I know, they'd be shocked'. As the years went on, he learned to cover up his problems and as a result he always had a bad feeling whenever mathematics cropped up – a combination of fear of being caught out and guilt at not properly facing up to it. He has a good job but his fear of mathematics regularly causes him problems. He might have bought this book in order to lay to rest the ghost of his mathematical failure once and for all.

While your name is unlikely to be Mel or Marti, maybe there is an aspect of their experiences of, and hopes for, mathematics learning that you can identify with.

Should I start at Chapter 01 and read right through the book?

The honest answer to this question is, 'It all depends . . .'. If your mathematics is reasonably sound, you might like to skip

Chapters 01 to 03, which deal with basic ideas of what a number is (hundreds, tens and units), how to add, subtract, multiply and divide and how to make a start with a simple four-function calculator. However, even if you do have a basic understanding of these things, you could always skim-read these chapters, if only to boost your confidence.

If, like Marti, you have an interest in helping someone else, say a child, then read these three chapters with a teacher's hat on. They should provide you with some ideas about how you can help someone else to grasp these ideas of basic arithmetic.

Can I succeed now if I failed at school?

There are a number of reasons that school mathematics may have been dull and hard to grasp. Here are some resources that you may have now which weren't available to you when you were 15.

Relevance

There were undoubtedly many more important things going on in most people's lives at the age of 8, 12 or 15 than learning about adding fractions or solving equations. Mathematics just doesn't seem to be particularly important or relevant at school. But as you get older, you are better able to appreciate some of the practical applications of mathematics and how the various mathematical ideas relate to the world of home and work. Even a willingness to entertain abstract ideas for their curiosity value alone may seem more attractive out of a school context. One student comments:

> Yes, I think you can associate more with it. When you get to our age, when you've a family and a home, I think you can associate more if you do put it to more practical things. You can see it better in your mind's eye.

Confidence

As an adult learner, you can take a mature approach to your study of mathematics and be honest in admitting when you don't understand. As one adult learner said after an unexpectedly exciting and successful mathematics lesson conducted in a small informal setting:

> I felt I could ask the sort of question that I wouldn't have dared ask at school: like 'What is a decimal?'

Motivation

A third factor in your favour now is motivation. Children are *required* to attend school and to turn up to their mathematics lessons whether they want to or not. In contrast, you have made a conscious choice to study this book, and the difference in motivation is crucial. Perhaps you have chosen to read the book because you need a better grasp of mathematics in order to be more effective at work. Or maybe you are a mathephobic parent who wants a better outlook for your own child. Or it is possible that you have always regretted that your mathematical understanding got lost somewhere, and it is simply time to lay that ghost to rest.

Maturity and experience

You are a very different person to when you were 8 or 15 years old. You now have a richer vocabulary and a wide experience of life, both of which will help you to grasp concepts you never understood before.

Hasn't mathematics changed since I was at school?

Mathematics *has* changed a bit since you were at school but not by as much as you think. The changes have occurred more in the language of mathematics than in the topics covered. Basic arithmetic is still central to primary mathematics and today's 12-year-olds are still having the same sort of problems with decimals and fractions as you did.

Will it help to use a calculator?

Yes. Most adults have a calculator, but perhaps many may rarely use them. This book should help you overcome any anxiety you may feel. As one student explains:

> It's only a thing with buttons, isn't it. All right, it takes you a while to know what each button is, but it's like driving a car. Once you've learnt which key to go to, you do it and that's it. It's just a case of learning. No, it didn't frighten me.

You will need to get a calculator before reading the following chapters. You won't need anything more sophisticated (or more expensive) than a simple 'four function' calculator (i.e. one which does the four functions of add, subtract, multiply and divide).

There are a number of reasons why this book has been written assuming a calculator is to hand and some of these are spelt out in more detail at the start of the next chapter. But two reasons to think about here are: Firstly, the calculator is more than simply a calculating device. As you work through this book, you will see how the calculator can be used as a means of learning and exploring mathematics. As one student commented:

> That was something that I enjoyed; knowing that you can divide and get larger numbers. I mean, that was a thing that, without the calculator, I would have been, . . . would never have believed, but because it was instant, it was something I could see straight away.

Secondly, being competent with mathematics isn't just some abstract skill that divides those who can from those who can't. Mathematical skills should be useful and insightful to you in your real world and in the real world that most of us inhabit.

This chapter ends with a section on children's mathematics. This theme will recur throughout the book and I hope that, even if you are not yourself a parent, you will find it interesting. My experience of working with adults has been that looking at children's errors and experiences in mathematics can be a fascinating 'way in' for the adults, giving a better understanding of the key ideas of mathematics.

Can I help my child?

Many parents who were unsuccessful at mathematics themselves have the depressing experience of watching their children follow in mother's or father's footsteps. Being hopeless at mathematics seems to run in families: or does it? While it certainly seems likely that *confidence* in mathematics passes from parent to child, it is less certain that 'mathematical ability' is inherited in this way. So if you consider yourself to be a duffer at mathematics, there is no reason to condemn your child to the same fate. Indeed, there is much that you can do to build your child's confidence and stimulate his or her interest in the subject.

Try to stimulate your child's thinking and curiosity about mathematical ideas. The ideal setting for these conversations might be in the kitchen, in a supermarket or on a long journey.

How, then, can parents encourage in their children a curiosity and excitement about mathematical ideas? Below are a few general pointers to indicate the sort of things you might say and do with your child to achieve this aim by creating what we professionals helpfully call a 'mathematically stimulating environment'.

- Where possible try to respond to your child's questions positively.
- Encourage the desire to have a go at an answer, even if the answer is wrong. Wrong answers should be opportunities for learning, not occasions for punishment or humiliation.
- Try to think about what makes children tick and the sort of things they might be curious about.
- Resist providing easy adult answers to your child's questions but rather try to draw further questions and theories from her/him.
- If a child gives a wrong answer it is probably for a good reason. Try to discover the cause of the problem – it could be, for example, that your question was not clearly expressed or that the child is simply not ready for the concept.
- There may be specific educational toys and apparatus which are helpful to have around. Having said that, a four-year-old will learn to count as effectively or ineffectively with pebbles or bottle tops as with 'proper' counting bricks!
- Finally, remember that 'how' and 'why' questions are at a higher level of curiosity than 'what' questions and these should be encouraged. (For example, *'why* does $2 \times 3 = 3 \times 2?'$ is a more stimulating question than 'what is $2 \times 3?'$)

The biggest barrier to learning mathematics is *fear*; the fear of being shown up as not understanding something which seems to be patently obvious to the rest of humanity. The best way of helping your child is to start by trying to overcome that fear in yourself. It doesn't matter if you don't know a simple fraction from a compound fracture; your competence at mathematics is less important than your willingness to be honest about what you don't understand yourself.

Just as with cooking, carpentry or playing the piano, learning mathematics doesn't happen just by reading a book about it. Although explanations can lead to understanding, practice is necessary if you are to achieve mastery.

02

the magic number machine

In this chapter you will learn:
- how to say 'hello' to your calculator
- about numbers and how they are represented
- how children first find out about numbers.

Numbers, numbers everywhere

Everywhere you look, numbers seem to leap out. They lie hidden in recipe books, are stamped onto coins and printed onto stamps, flash up on supermarket check-outs, provide us with breakfast reading on cornflakes packets, are displayed on buses, spin round on petrol pump dials . . . . It seems that, whatever task we want to perform, numbers have some role to play. Here are few more examples.

- A farmer will check that all the cows are in by *counting* them as they go through the gate.

- At church, the vicar reads out the hymn number. Members of the congregation are able to find the hymn in their hymn books only if they know how *numbers are organized in sequence*.

- The recipe book says 'pour into a 9-inch square tin and bake at 180°C for 45–50 minutes'. For this to make sense, we need to know about *measurement* of size in inches, temperature in degrees Celsius and time in minutes.

- Competitive sport is totally based on numbers, usually in the form of *scoring* – half-time scores, number of runs, highest break, winning times . . . .

You can probably think of lots more examples of your own. It's hard to imagine a world with no numbers. How would we survive without them? What alternative ways might we think up to organize human activity if we couldn't count on numbers to control them?

Exercise 2.1 is designed to help you to be more aware of the role that numbers play in your life.

EXERCISE 2.1 How long can you last without numbers?

Take a waking period of, say, 30 minutes of your life and see how many situations require some use and understanding of numbers. Then read on.

I tried the exercise while writing this chapter on my computer. Within just a few seconds I found that I was adjusting the line *width of the text* to exactly sixteen and a half centimetres.

Minutes later the 'phone rang and I was aware that I stated my *telephone number* on picking up the receiver.

The 'phone call was to fix a meeting, so I had to confirm *date* and *time* in my diary.

Shortly afterwards I went to make a cup of coffee (I don't have a very long attention span!). This involved putting milk and water into a cup, placing it into the microwave and keying 222 (2 minutes and 22 seconds) into the *timer*.

Numbers are the building bricks of mathematics. So, just like being able to read, there are certain basic mathematical skills that you need in order to live a normal life. For example, being able to:

- read numbers and count
- tell the time
- handle money when shopping
- weigh and measure
- understand timetables and simple graphs.

But as well as having a useful, practical side, solving problems with mathematics can also be challenging and fun. Anyone who has done dressmaking or carpentry knows that mathematics can be used in two ways. One is in the practical sense of measuring and using patterns and diagrams. The other is more abstract – 'How can I cut out my pattern so as to use up the least material?', 'Can I use the symmetry of the garment to make two cuttings in one?' The pleasure you get from solving these sorts of problems is what has kept mathematicians going for four thousand years!

Most of all, mathematics is a powerful tool for expressing and communicating ideas. Sadly, too few people ever get a sense of the 'power to explain' that mathematics offers.

Let us return now to these building bricks, the numbers, and see what sense your calculator makes of them.

Saying hello to your calculator

Most basic calculators look something like this (see over).

display screen

the 'off' key — OFF

√

%

÷ — the 'divide' key

MRC

M−

M+

× — the 'multiply' key

7

8

9

− — the 'subtract' key

the number pad

4

5

6

+ — the 'add' key

1

2

3

= — the 'equals' key

the 'on' and 'clear' key — ON/C

0

•

zero

the 'decimal point'

Before we tackle the hard stuff, you should start by saying hello to your calculator. Usual etiquette is as follows.

- Press the 'On' switch (probably marked $\boxed{ON}$ or $\boxed{ON/C}$)
- Now key in the number 0.7734
- Turn your calculator upside down and read its response on the screen.

Good! Having now exchanged 'hELLOs', this is clearly the basis for a good working relationship!

As was explained in the previous chapter, this book aims to teach you basic mathematics *with a calculator*. There are a number of good reasons for presenting the book in this way, some of which are listed below.

- For some types of problem, calculators take the slog out of the arithmetic and allow you to focus your attention on understanding what the problem is about.

- Examination boards have largely designed their examinations on the assumption that all candidates have access to a calculator.
- Most importantly, the calculator provides a powerful aid to learning mathematics. See if you are more convinced about this by the time you have finished this book.

Introducing the counting numbers

Let us begin at the beginning; with the counting numbers 1, 2, 3, 4, 5, . . .

No doubt they look familiar enough. But before you reach for pencil and paper, let's see how they look on your calculator.

EXERCISE 2.2 Entering numbers

Switch on your calculator and key in the following sequence

1 2 3 4 5 6 7 8 9

Look carefully at the screen.

Write down what is recorded and make a note of exactly how the numbers are displayed.

If you look carefully at each number, you should see that it is made up of a series of little dashes. For example, the three requires five dashes and is written as follows.

You might like to consider in the next exercise which of the numbers from 0 to 9 requires the fewest and which the most dashes to be displayed.

EXERCISE 2.3 Displaying numbers

a Check how the other numbers are represented and count the number of dashes each number needs. Then fill your answer into the table below (one, the 3, has already been done for you).

NUMBER	0	1	2	3	4	5	6	7	8	9
DASHES				5						

b See if you can sketch any new combinations of dashes that are not already covered by the numbers. If you were a calculator designer, could you turn these into anything useful?

Your table should now enable you to answer the question posed earlier, namely to spot which number requires the fewest and which the most dashes to be displayed. Was your earlier guess correct? In fact, the number 1 uses least dashes (two), while the 8 uses most (all seven).

There are a few combinations not covered by the numbers. For example:

and

One possible application of these shapes is to spell letters. For example, the 'E' above could be used to refer to 'Error' if an impossible key sequence was pressed. The 'C' could perhaps be used to mean 'Constant operating' or maybe 'Careful, you're hitting my keys too hard!'

Ordering numbers

Most calculators can be set up to produce number sequences. Here is a simple one to try.

Press ⊞ 1

Now press ⊟ repeatedly.

You should see the counting numbers in sequence: 2, 3, 4, 5, 6, . . .

If this has not happened, try the following.

Either: Press   1 ⊞ ⊞ and then the ⊟ repeatedly.

Or: Press   1 ⊞ 1 and then the ⊟ repeatedly.

What is going on here is that your calculator is doing a constant 'add 1' calculation. The capability to perform 'constant' calculations like this is available on most calculators in some form. It is an extremely useful feature and exactly how and when it can be used will be explained later.

Tens and units

Let's have another look at the sequence of counting numbers.

Repeat the instruction above, using the constant to add 1. As before, repeatedly press ⊟ until the calculator displays 9. The display should look like this.

> 9

Pause for a moment and then press the ⊟ once more.

You should now see the following.

> 10

Most people will recognize that this is simply a ten, but note that it is quite different from the previous numbers. Focus on the fact that there are now not one but two figures displayed here. What has happened in the move from nine to ten is that the nine has changed to a zero and a new figure, the 1, has appeared to the left of the zero, thus:

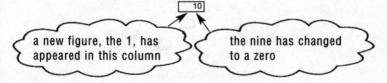

a new figure, the 1, has appeared in this column

the nine has changed to a zero

It is worth reflecting on the fact that numbers do not need to do this. As you saw from Calculator Exercise 1.2, the calculator has a few more squiggles up its sleeve which might be used for extra numbers. For example, the numbers could be extended to look like this:

zero	one	two	three	four	five	six
0	1	2	3	4	5	6

seven	eight	nine	ten	eleven	twelve	thirteen

But, as it happens, our number system doesn't look like this. Probably for the reason that most humans were born with ten fingers and ten toes, we have come to an agreement that only ten unique characters are needed for counting. These are 0, 1, 2, 3, 4, 5, 6, 7, 8 and 9.

These ten characters are called the *numerals*. The word numeral refers to how we write numbers, rather than being concerned with how many things the numbers represent.

After counting as far as 9, we simply group numbers in tens, and count how many tens and how many left over. For example, here is a scattering of just some of the one pound coins that I happened to find when tidying up my daughter's money box.

In order to count them, I might group them in tens, as follows.

So the columns represent the tens, and the ones are the units. There are therefore twenty three coins here: two tens and three units.

If you feel that you need practice at dealing with tens and units for slightly larger numbers, have a go now at Exercise 2.4.

EXERCISE 2.4 Practising with tens and units

Switch your calculator on and once again set up the constant 'add 1' by pressing the sequence:

+ 1 = (or by another method, depending on your particular calculator.)

Now enter the number 37 and press the = key four times.

The result should be 41.

In other words,

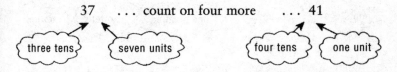

Note that pressing any key at this point other than the number keys and = is likely to destroy the constant setting. If, for example, you have already pressed the 'clear' key, probably marked C or ON/C, you impetuous person, you will need to reset the constant by re-keying + 1 = or an equivalent key sequence.

Once again, at the risk of being boring, you are reminded not to clear the screen after each sequence, as this will destroy the constant setting.

Enter	Press	Expected tens	Expected units	Answer
44	= = =	4	7	47
59	= = = = =	6	4	64
73	= = = = = = = =	8	1	81
66	= = = =	7	0	70

Don't press 'Clear' here

This idea of grouping in tens is the very basis of counting up as far as 99. The next section goes beyond tens and units, into the world of hundreds and thousands.

Hundreds, thousands and beyond

EXERCISE 2.5 Handling hundreds

Switch your calculator on and once again set up the constant 'add 1' by a method suitable for your particular machine.

Now enter the number 96 and press the ⌐=⌐ key four times.

The result should be 100.

In other words,

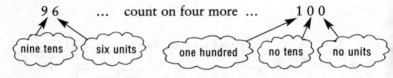

You should not press any key other than the number keys and ⌐=⌐. As before, if you have destroyed the constant, you will need to reset it by re-keying ⌐+⌐ 1 ⌐=⌐ or an equivalent key sequence.

Remember also not to clear the screen after each sequence.

Enter	Press	Expected hundreds	Expected tens	Expected units	Answer
198	⌐=⌐ ⌐=⌐ ⌐=⌐	2	0	1	201
399	⌐=⌐ ⌐=⌐ ⌐=⌐ ⌐=⌐ ⌐=⌐	4	0	4	404
696	⌐=⌐ ⌐=⌐ ⌐=⌐ ⌐=⌐ ⌐=⌐ ⌐=⌐	7	0	2	702
897	⌐=⌐ sixteen times	9	1	3	913

The largest number you can produce with three figures is nine hundred and ninety-nine. If you want to get any larger, you need to regroup and create a new category of 'ten hundreds', which we call one thousand.

Let's explore thousands by setting the calculator to count up in much larger steps; this time in intervals one hundred at a time. You may be able to work out for yourself how to do this. If not, try one of the following sequences.

Press [+] 100 and then the [=] repeatedly.
Or: Press 100 [+] [+] and then the [=] repeatedly.
Or: Press 100 [+] 100 and then the [=] repeatedly.

As you do this exercise, try to develop a sense of what to expect at the next press of [=], particularly when the number in the hundreds column is a 9.

To end this section, it is useful to be able to say numbers as well as to write and interpret them.

EXERCISE 2.6 Saying numbers out loud

Set your calculator constant to add 1423.

This is done by pressing [+] 1423 (or 1423 [+] [+] or 1423 [+] 1423).

Now, each time you press the [=] key, try to say out loud the number you see on the screen. Incidentally, it helps if you say the numbers VERY LOUDLY INDEED. Don't worry if other members of your household or your dog think you are crossing the final frontier. This is normal behaviour for mathematical geniuses.

When you have completed this exercise, you can check your answers with those given at the end of the chapter.

The number system represents a central theme in the work of teachers of children in the early years of schooling. You may have children of your own or may be interested in how these ideas are tackled with young children. The final section of this chapter invites you into the infant classroom to see what things go on and what sorts of notions and difficulties young children have with number.

EXERCISE 2.7 Ordering your figures

Using the figures 3, 9, 4 and 6 once only, write down:

a the largest possible four figure number
b the smallest possible four figure whole number.

Beyond thousands and tens of thousands lie two additional words that you should know: millions and billions.

A million is a thousand thousand or 1 000 000.

A billion is a thousand million or 1 000 000 000.

Incidentally, notice how, with very large numbers like these, the digits are often grouped in threes simply to make them easier to read.

To end this section, the term *place value* needs to be explained. Place value is really another way of describing the whole idea of hundreds, tens and units. It refers to the key idea of how our number system works, namely that the 'place' of a digit (in other words, its position in the number) is what determines its value. For example, the two-digit number 37 is written as 3, followed on the right by 7. By convention, we agree that the first of these digits, the three, refers to three tens while the seven refers to seven units. To understand this point is to understand the principle of 'place value'.

Children and numbers

For most children, number words first come into their world through songs ('Five little speckled frogs sitting on a log', 'One, two, three, four, five, once I caught a fish alive', and so on). A feature of many such songs is that the numbers are sung in sequence – sometimes the numbers go up and sometimes they go down. This property of the way that numbers follow on from each other in sequence 1, 2, 3, 4, 5... is the *ordinal* property of numbers (ordinal as in the 'order').

The *number line* shown below is a helpful way of enabling children to form a mental picture of the sequence of numbers.

However, there is quite a step for children from understanding numbers in a number line, to becoming aware of what the three-ness of three really means in terms of the number of objects – 3 toys, 3 bricks, 3 sweets, 3 calculators, and so on. It is a major breakthrough for a child when she learns that 'three' is a description that can be applied to a whole variety of different collections (or sets) of things. She will need many occasions in which she physically grasps three objects before three-ness becomes a concept that she can grasp mentally. This idea of a number describing *how many things there are* is known as the *cardinal* property of numbers. Many school activities in the early years deal with this concept by means of play involving counting out various objects – bricks, bottle tops, match boxes, and so on.

Finally, just in case you thought that helping children to understand number was a straightforward exercise, try to pick your way through the following exchange between a four-year-old and his teacher, recorded in the book *Wally's Stories* by Vivian Paley.

'We have three 12s in this room', Wally said one day. 'A round 12, a long 12, and a short 12. The round 12 is the boss of the clock, the long 12 is the ruler, and the short 12 is on a calendar.'

'Why is the 12 on the calendar a short 12?', I asked.

'Me and Eddie measured it. It's really a five. It comes out five on the ruler.'

'Right. It's five.' Wally stared thoughtfully at the clock.

'I'm like the boss of March because my birthday is March 12. The 12 is on the top of the clock.'

Summary

In this chapter you were introduced to your calculator and shown how to use its constant facility. You were then shown how to use the *calculator constant* to count in ones or in any interval, and you were then asked to count your way through our whole number system, based on *tens, hundreds* and *thousands*. By closely examining how numbers are represented on the calculator display, we looked at the *numerals*, i.e. how numbers are written. But numbers have other features worth exploring. Firstly, they form a natural sequence. This is known as their *ordinal* property and is nicely represented on a *number line*. They are also important as a way of describing 'how many', which is their *cardinal* property.

Comments on exercises

Exercises 2.1 to 2.5	no additional comments
2.6	Saying numbers out loud

Written on the screen	*Said VERY LOUDLY INDEED*
1423	one thousand four hundred and twenty-three
2846	two thousand eight hundred and forty-six
4269	four thousand two hundred and sixty-nine
5692	five thousand six hundred and ninety-two
7115	seven thousand one hundred and fifteen
8538	eight thousand five hundred and thirty-eight
9961	nine thousand nine hundred and sixty-one
11384	eleven thousand three hundred and eighty-four
12807	twelve thousand eight hundred and seven

2.7 **a** the largest possible four figure whole number is 9643
b the smallest possible four figure whole number is 3469.

03

calculating with numbers

In this chapter you will learn:
- some important number words – like prime, square, odd and even
- what calculations you need in different situations
- about the 'four rules': add, subtract, multiply and divide
- how negative (i.e. minus) numbers fit in.

Note: In this and all subsequent chapters, comments on exercises are given at the end of the chapter.

Properties of numbers

Before tackling the sorts of calculation that we normally do with numbers, it is worth looking at some of the more useful properties of numbers. You will find out what is meant by *odd, even, prime, rectangular* and *square* numbers. These terms are best understood by seeing the numbers arranged in patterns on the table in front of you. So, before reading on, try to get hold of ten or twelve small identical objects (coins, buttons, or paper clips, for example).

Even or odd

Choose a selection of your objects (any number between 1 and 12 will do) and try to arrange them into two rows, like this:

If, like me, you chose a number of coins that produced two equal rows, then your selection was an even number of coins. In this case, choosing ten coins produced two even rows of five each. So ten is an even number.

However, perhaps your selection didn't work out like this and, when the two equal rows were formed, there was an odd one over – like this:

the odd one over

When any selection of things that are laid out into two rows produce an odd one over, then there must have been an odd number of them. In this case, choosing eleven coins produced two even rows of five each plus an odd one over. So eleven is an odd number.

Exercise 3.1 will give you practice at deciding whether a number is even or odd.

EXERCISE 3.1 Even or odd?

For each number, mark the box corresponding to whether it is even or odd. The first one has been done for you.

Number	11	7	2	12	8	6	3	1	0
Even	☐	☐	☐	☐	☐	☐	☐	☐	☐
Odd	☒	☐	☐	☐	☐	☐	☐	☐	☐

Prime, rectangular and square

Now choose six of the coins. Notice that they can be arranged in a rectangle as two rows of three, like this:

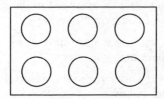

or as three rows of two, like this:

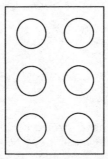

Either way, six is a rectangular number because it *can* be arranged in the form of a rectangle. Similarly, the number 18 is rectangular because 18 coins could be arranged like this.

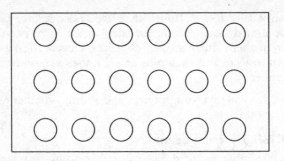

Now try the same task with seven coins. You will soon find that this is impossible. No matter how you move them around, the coins will not form a rectangle; they can only be placed on a line, like this.

So seven is not a rectangular number.

Any number which can't be arranged in the shape of a rectangle is called *prime*, so 7 is a prime number.

Pause for a few moments now and think about what other prime numbers there are.

Now choose 9 coins and arrange them into three rows and three columns, as shown below.

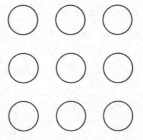

Notice that the coins have formed a square shape. The number 9 is a square number because it *can* be arranged in the form of a square. What other square numbers can you think of?

The next exercise will give you the opportunity to practise your understanding of these three terms, prime, rectangular and square numbers.

EXERCISE 3.2 Prime, rectangular or square?

For each number, mark the boxes corresponding to whether it is prime, rectangular or square. The first one has been done for you. Note that 9 is both rectangular and square, so two boxes have been marked.

Number	9	7	2	12	8	5	3	4	11
Prime	☐	☐	☐	☐	☐	☐	☐	☐	☐
Rectangular	X	☐	☐	☐	☐	☐	☐	☐	☐
Square	X	☐	☐	☐	☐	☐	☐	☐	☐

THAT NUMBER SEVEN, SHE'S OUR PRIME SUSPECT!

You have already been introduced to the idea of a square. You should also know that the opposite of a square is called a *square root*, and it is written as $\sqrt{}$. A few examples should illustrate what square root means.

The square of 5, written as 5^2, is 25.
The square root of 25, written as $\sqrt{25}$ is 5.

The square of 9, written as 9^2, is 81.
The square root of 81, written as $\sqrt{81}$ is 9.

The square of 10, written as 10^2, is 100.
The square root of 100, written as $\sqrt{100}$ is 10.

What are the 'four rules'?

The four most basic things you can do to any two numbers is add (+), subtract (−), multiply (×) or divide (÷) them. These are known as the four rules (sometimes called the four 'operations').

I've set out below some simple examples explaining what each of these four rules means. Keep your coins handy and, if you think it helpful, act out the various calculations using the coins. Even though you may feel a bit daft doing this, you will find that this will help you to remember it later. (Having said that, I have found that when I have been a bit daft, other people seem to remember it later.)

Adding

Here is a simple example of an adding situation.

I start with 2 gold rings

... and my true love gives me 3 more

How many gold rings do I have altogether?

This calculation would be entered in your calculator as follows.

- First, switch on.
- Then press 2 ⊞ 3 ⊟
  and the result, 5, appears on the display.

If you feel you need some more exercises on the rings, use your calculator to do Exercise 3.3 now.

EXERCISE 3.3 Give me a ring

a You have 13 gold rings and your true love gives you 13 more. How many do you have altogether?

b You have 9 gold rings and your true love gives you 45 more. How many do you have altogether?

c You have 29 gold rings and your true love gives you 83 more. How many do you have altogether?

d You have 316 gold rings and your true love gives you 477 more. How many do you have altogether?

Subtracting

Here is a simple example of a subtracting situation.

You start with 8 apples   🍎🍎🍎🍎🍎🍎🍎🍎

and generously give 5 away   🍎🍎🍎 → 🍎🍎🍎🍎🍎

How many apples do you have left?   🍎🍎🍎

This calculation would be entered in your calculator as follows.

- First, switch on.
- Then press 8 ⊟ 5 ⊟ The result, 3, appears on the display.

If you feel you need some more practice at giving away apples, use your calculator to do Exercise 3.4 now.

EXERCISE 3.4 Apple folly!

a You have 13 apples and you give away 4.
How many do you have left?
b You have 42 apples and you give away 17.
How many do you have left?
c You have 89 apples and you generously give away 53.
How many do you have left?
d You have 277 apples and you foolishly give away 218.
How many do you have left?

Before we leave adding and subtracting (to move on to multiplying and dividing), have a look at the diagram below, which illustrates a *number line*. This is a useful way of representing numbers and the operations of addition and subtraction. The number line itself is nothing more than a straight line with the numbers placed in sequence on it. The diagram below has been drawn to show the calculation 8 – 3. The procedure is to start at the value 8 (on the right hand arrow, labelled 'Start'), take three steps to the left, and the answer is the point on the number line where you stop. In this example, this is at the value 5.

So 8 – 3 = 5.

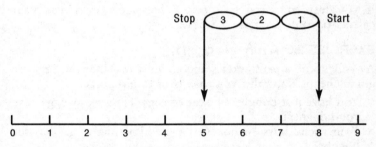

a number line showing the calculation 8 – 3

In general, calculations on the number line are organized as follows.
• Additions involve taking steps to the right.
• Subtractions involve taking steps to the left.

You could spend a few minutes now checking that you can represent simple calculations like this one on a number line.

Multiplying

An egg-box contains six eggs.

You buy three boxes. How many eggs do you have altogether?

In this eggs-sample you found three lots of six things. This is entered into the calculator as follows:

- Press 3 ☒ 6 ☲ and the result, 18, appears on the display.

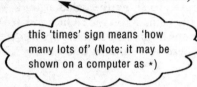

this 'times' sign means 'how many lots of' (Note: it may be shown on a computer as *)

If you feel you need to crack a few more eggs, use your calculator to do Exercise 3.5 now.

EXERCISE 3.5 A run on eggs (1)

At your local supermarket someone tells you that the eggs are going cheap. Naturally, you rush over to buy some.

a You have just bought 4 boxes of eggs. How many eggs have you bought?
b Your friend buys 18 boxes of eggs. How many eggs has she bought?
c A bakery buys 114 boxes of eggs. How many eggs have they bought?
d That day, the supermarket sells 1673 boxes of eggs, all going cheap. How many eggs have they sold?

Dividing

Division is the fourth and most troublesome of the four rules. Here is another typical everyday scenario where division might crop up.

Esther keeps chickens in her back garden. On a Monday, she collects 30 eggs. How many boxes of eggs will she need?

Esther's 30 eggs

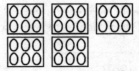

Each egg-box holds six eggs

How many egg-boxes will she need?

This is entered into the calculator as follows:

Press 30 ⌈÷⌉ 6 ⌈=⌉ and the result, 5, appears on the display.

> this 'divide' sign means 'shared by' (*Note*: It may be shown on a computer as /)

If you feel you need to box a few more eggs, use your calculator to do Exercise 3.6 now.

EXERCISE 3.6 A run on eggs (2)

a On a Monday, she collects 48 eggs. How many egg-boxes will she need?
b On Tuesday, she collects 42 eggs. How many egg-boxes will she need?
c Over a week, she collects 258 eggs. How many egg-boxes will she need?
d An egg packing station collects 15 732 eggs. If they pack them into boxes of 18, how many egg-boxes will they need?

If you have just tackled Exercise 3.6, you may have noticed that the number of eggs were deliberately fixed so that the division worked out exactly. In other words, in every case, there were the correct number of eggs to fill each box completely, with no spare eggs left over. In the real world, of course, calculations rarely work out so neatly. What to do when division doesn't work out exactly is discussed in Chapters 04 and 05.

Knowing what sum to do

What causes some confusion when using the four rules is that different people use different words to describe them. Most of these terms are listed in Exercise 3.7. See how many you recognize.

EXERCISE 3.7 The terms used to describe the four rules

Tick which term refers to which rule.

Term	+	−	×	÷
add				
and				
difference				
divide				
from				
goes into				
how many more				
how many less				
less				
minus				
multiply				
plus				
product				
share				
sum				
subtract				
take away				
times				

With a calculator to hand, actually doing a calculation is usually straightforward. The real skills are knowing what sum to do and know how to interpret your answer. Here is an illustration.

EXAMPLE 1

Calculate $5 \times 4 + 7$

Solution

The calculator sequence is 5 $\boxed{\times}$ 4 $\boxed{+}$ 7 $\boxed{=}$ giving the answer, 27.

EXAMPLE 2

I have four bottles of milk delivered each weekday and seven at the weekend. How many bottles are delivered in a week?

Solution

The calculation is made up of two parts.

Five *lots of* 4 bottles (one for each of the five weekdays) is a multiplication: 5×4

and seven at the weekend is an addition: $+ 7$

So the solution, as before, is 5 $\boxed{\times}$ 4 $\boxed{+}$ 7 $\boxed{=}$

Giving the answer, 27 bottles.

Here are some for you to try yourself.

EXERCISE 3.8 Calculations in context

a A bus sets out from the depot with 27 people.
  Calculate the number of people on the bus, if:
  (i) after the first stop, 18 people get on and 9 get off.
  (ii) after the second stop, 12 people get on and 21 get off.
  (iii) after the third stop, 5 people get on and 16 get off.

b Over eight minutes, the oven temperature rose from 21 degrees Celsius to 205 degrees Celsius.
  (i) What was the temperature rise?
  (ii) What was the average temperature rise per minute?

c Denise drinks 15 glasses of water per day. How many will she have drunk:
  (i) in a week?
  (ii) during the month of July?
  (iii) in a year?

d A bottle of wine is said to provide enough for about 7 glasses. If a glass holds about 10 cl, how much does a bottle hold?

Doing sums on paper

In this section you are asked to put your calculator to one side and look at how calculations were done before calculators were invented. No doubt you spent many hours at school adding, subtracting, multiplying and dividing. Here is a summary of the methods as they are currently taught in most schools. The solutions have been broken down into stages to help you follow what is going on.

Adding with pencil and paper

EXAMPLE 1

Calculate   27 + 68

Solution

Stage 1	**Stage 2**	**Stage 3**
Set out the sum like this, with units under units and tens under tens	Add the units	Write the 5 in the units column ...

tens units	7	
2 7	+ <u>8</u>	2 7
+ <u>6 8</u>	1 5	+ <u>6 8</u>
	or 1 ten 5 units	5

Stage 4	**Stage 5**
... and write the 1 in the tens column	Add the tens 2 + 6 + 1 = 9 and write the 9 in the tens column

2 7	2 7
+ <u>6₁8</u>	+ <u>6₁8</u>
5	9 5

EXAMPLE 2

Calculate   173 + 269

Solution

Stage 1	**Stage 2**	**Stage 3**
Set out the sum like this, with units (U), tens (T) and hundreds (H) in their columns	Add the units 3 + 9 = 12	1 7 3 + <u>2 6 9</u> 2

H T U 1 7 3 + <u>2 6 9</u>	3 + <u>9</u> 1 2 1 ten 2 units	Write the 2 in the units column ...

Stage 4	**Stage 5**	**Stage 6**
... and write the 1 in the tens column	Add the tens 7 + 6 + 1 = 14 14 tens is 1 hundred and 4 tens	Write the 4 in the tens column ...

1 7 3
+ <u>2 6,9</u>
    2
	1 7 3
	+ <u>2 6,9</u>
	    2
		1 7 3
		+ <u>2 6,9</u>
		   4 2

Stage 7	**Stage 8**	**Stage 9**

... and write the 1 in the hundreds column
1 7 3
+ <u>2,6 9</u>
  4 2
	Add the hundreds
	1 + 2 + 1 = 4
	1 7 3
	+ <u>2,6 9</u>
	  4 2
		Write the 4 in the hundreds column
		1 7 3
		+ <u>2,6 9</u>
		4 4 2

EXERCISE 3.9 Practising adding with pencil and paper

Try these addition sums now. You can check your answers using your calculator.

a 43 + 67
b 162 + 77
c 437 + 239
d The sum of 259 and 952

Subtracting with pencil and paper

EXAMPLE 1

Calculate 98 – 45
Solution

Stage 1	**Stage 2**	**Stage 3**
Set out the sum like this, with units under units and tens under tens	Subtract the units	Write the 3 in the units column …

Stage 1

Set out the sum like this,
with units under units
and tens under tens

tens units
  9 8
– 4 5

Stage 2

Subtract the units

  8
– 5

8 – 5 = 3

Stage 3

Write the 3 in the
units column …

  9 8
– 4 5
    3

Stage 4

Subtract the tens

  9 8
– 4 5
    3
9 – 4 = 5

Stage 5

Write the 5 in the
tens column

  9 8
– 4 5
  5 3

This first example was carefully chosen so that, in both the tens
and in the units column, you were subtracting a smaller number
from a bigger number. The next example shows what to do if,
at any point in the calculation, you have to take bigger from
smaller.

EXAMPLE 2

Subtract 27 from 63

This time, you are asked to tackle the problem from your
imagination, rather than by following a set of rules.

Imagine that you are a shop-keeper selling tins of beans. The tins
come packed in cases, ten tins to a case. On the floor behind the
counter are six unopened cases. A seventh case is open on the
counter, with just three tins in it. So your current stock of beans is
63 tins (six lots of tens and three singles).

Suddenly, a customer walks in – clearly a big spender – and asks for 27 tins of beans. Starting with the seven tins, you have no choice but to open up one of the cases. You now have 10 + 3 = 13 loose tins, and you count out 7 of them into the customer's trolley. Then you hand over two cases (two lots of tens), making 27 in all. What you have left are three unopened cases and six loose tins. Your stock is now 36 tins.

This approach to subtraction is called the method of 'decomposition', because it involves 'decomposing' or 'breaking up' lots of tens (or lots of hundreds, thousands, and so on). Remember, decomposition is only necessary when you are trying to take bigger from smaller and you have to go and open up a bigger case.

EXAMPLE 3

Taking 48 coins from 95 coins

Stage 1   Set the coins out in columns, with ten coins in each column, thus:

Ninety five coins

9 tens   and 5 units

Stage 2   Take away 8 coins

Unfortunately there are only five loose coins, so we cannot take 8 away directly. We need to break up one of the columns of ten into loose coins, making a total of 8 tens and 15 units thus:

8 tens   and 15 units

You can now take away the 8 units.

$15 - 8 = 7$

This leaves 8 tens and 7 units, thus:

8 tens   and 7 units

Stage 3   Take away the 40 coins

You now take away 4 lots of ten from the remaining 8 lots of ten, leaving 4 lots of ten.

So the answer is 47:

i.e. 4 lots of ten and 7 units.

Here is how the decomposition process would look on paper.

EXAMPLE 4

Calculate on paper 95 – 48

Solution

Stage 1	Stage 2	Stage 3
Set out the sum like this, with tens (T) and units (U)	Subtract the units	Decompose one of the 9 tens to get 8 tens and 15 units
T U 9 5 – 4 8	5 – 8 5 – 8 = ? Can't take bigger from smaller	8 ¹5 – 4 8

Stage 4	Stage 5
Subtract the units	Subtract the tens
	$8 - 4 = 4$
$15 - 8 = 7$	Write the answer
Write the answer into	into the tens
the units column	column

```
   8 ¹5          8 ¹5
 -  4 8        -  4 8
      7           4 7
```

This confirms the answer we got before, 47.

Exactly the same approach applies when you are subtracting numbers of hundreds, thousands and so on, as the next example shows.

EXAMPLE 5

Calculate on paper   442 − 173

Solution

Stage 1	Stage 2	Stage 3
Set out the sum like this, with units (U), tens (T) and hundreds (H)	Subtract the units	Decompose one of the 4 tens to get 3 tens and 12 units

```
   H T U         2            4 3¹2
   4 4 2       - 3          -  1 7 3
 - 1 7 3
```

Stage 2: $2 - 3 = ?$
Can't take bigger from smaller

Stage 4	Stage 5	Stage 6
Subtract the units	Subtract the tens	Decompose one of
$12 - 3 = 9$		the 4 hundreds to
Write the answer in the		get 3 hundreds and
units column		13 tens

```
   4 3¹2          4 3¹2         3¹3¹2
 - 1 7 3        - 1 7 3       - 1 7 3
       9              9             9
```

Stage 5: $3 - 7 = ?$
Again, can't take bigger from smaller

Stage 7	Stage 8	Stage 9
Subtract the tens	Subtract the	Write the 2 in the
13 – 7 = 6	hundreds	hundreds column
Write the answer in the	3 ¹3 ¹2	
tens column	– 1 7 3	3 ¹3 ¹2
3 ¹3 ¹2	6 9	– 1 7 3
= 1 7 3		2 6 9
6 9	3 – 1 = 2	

EXERCISE 3.10 Practising subtracting with pencil and paper

Try these subtraction sums now. You can check your answers using your calculator.

a 67 – 43
b 162 – 77
c 437 – 239
d The difference between 259 and 952

How do the four rules connect?

The four rules are, of course, closely connected to each other. Exercise 3.11 is designed to let you 'discover' some of these connections for yourself. It is best done with the help of a calculator, though this is not essential.

EXERCISE 3.11 Connecting the four rules

Fill in the blanks in the sequences below.
(Those blanks marked in a long rectangle, thus, ⬭▭⬭, refer to a number that results from a calculation. Blanks marked ▭ refer to a 'rule' like + or –.)

a After completing the blanks, jot down on the right-hand side under 'Comment' what you think your result suggests about the four rules.

Key sequence   *Comment*

3 ⊞ 2 ⊟ [   ] ⊟ 2 ⊟ [   ]
4 ⊞ 3 ⊟ [   ] [ ] 3 ⊟ [4]
3 ⊠ 2 ⊟ [   ] [ ] 2 ⊟ [3]
4 ⊠ 3 ⊟ [   ] [ ] 3 ⊟ [4]
20 ⊟ 5 ⊟ [   ] [ ] 5 ⊟ [20]

b 3 ⊞ 3 ⊞ 3 ⊞ 3 ⊟ ☐

How many lots of 3 did you add? ☐

4 ⊠ 3 ⊟ ☐

c 6 ⊟ 2 ⊟ 2 ⊟ 2 ⊟ ☐

6 ⊡ 2 ⊟ ☐

Negative numbers

You may have noticed that the subtraction sums which you have been asked to do so far have been artificially 'set up' so that you have been taking smaller from larger. Basically, subtraction is normally seen very much in terms of 'taking away' objects, and so, if you start off with three objects, you can't take more than three away. However, subtraction doesn't always involve moving objects around. Look at these two examples:

- Although I had only £3 in my bank account, I wrote a cheque for £5.
- The temperature was 3°C and it dropped a further five degrees overnight.

The banking system hasn't ground to a halt or the thermometer exploded as a result of these events. We simply solve the problem by inventing a new set of numbers less than 0. In bank statements these have the letters O/D (standing for overdrawn) beside them. Usually, however, we just call them *minus*, or *negative* numbers. So really the number line should be extended to the left to look like this.

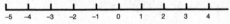

number line showing negative numbers

Subtracting 5 from 3 can be shown on the number line as follows:

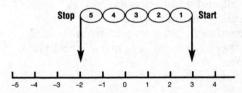

Starting at the number 3 (the right-hand arrow), take five steps to the left, taking us to the answer, –2, said as 'minus two'.

You can check this result by pressing the corresponding key sequence on your calculator. 3 ⌐−⌐ 5 ⌐=⌐

Now have a go at the practice exercise below, which should help to consolidate some of the key points of the chapter.

EXERCISE 3.12 Practice exercise

1 Set yourself a few simple 'sums' using the four rules of +, −, × and ÷. Check your answers using a calculator. Do the 'sums' involving + and − again by drawing a number line and moving, respectively, right or left. Check that you get the same answer as with the calculator.

2 The place value of the 6 in the number 365 is <u>ten</u>.
What is the place value of the digit 6 in the following numbers?

Number	365	614	496	16 042	1 093 461
Place value	ten				

3 7°C is three degrees less than 10°C.
Find the temperature which is three degrees less than the following:

Temperature °C	10	4	21	−6	−10	0	3	−3
Three degrees less	7							

4 Does it matter in what order you add, subtract, multiply and divide numbers?
For example, does 23 ⌐×⌐ 15 ⌐=⌐ give the same answer as 15 ⌐×⌐ 23 ⌐=⌐?
Use your calculator to explore.

5 (i) Does the square of an even number always give an even number?
(ii) Does the square of an odd number always give an odd number?
(iii) Are all odd numbers prime?
(iv) Are all prime numbers odd?
(v) Write down the numbers from 1 to 20, and indicate whether each number is prime, rectangular, odd, even or square.

Summary

This chapter started by examining some of the properties of numbers – whether they are even or odd, prime, rectangular or square, for example. Next, the so-called 'four rules' of add (+), subtract (−), multiply (×) and divide (÷) were explained, both on paper (in the Answers to Exercise 3.7) and with the help of a calculator. You were shown how to use a 'number line' to represent numbers and simple calculations. The language of arithmetic is an area which causes confusion and words like *sum* and *product* were defined. The four rules are, of course, closely connected to each other and these interconnections were explored with the help of a calculator. Finally, you were introduced to negative numbers; these can be thought of as the numbers that appear to the left of zero on the number line.

Answers to exercises for Chapter 03

3.1

Number	11	7	2	12	8	6	3	1	0
Even			×	×	×	×			×
Odd	×	×					×	×	

3.2

Number	9	7	2	12	8	5	3	4	11
Prime		×	×			×	×		×
Rectangular	×			×	×			×	
Square	×							×	

3.3

a $13 + 13 = 26$ rings
b $9 + 45 = 54$ rings
c $29 + 83 = 112$ rings
d $316 + 477 = 793$ rings

3.4

a $13 − 4 = 9$ apples
b $42 − 17 = 25$ apples
c $89 − 53 = 36$ apples
d $277 − 218 = 59$ apples

3.5

a $6 × 4 = 24$ eggs
b $6 × 18 = 108$ eggs
c $6 × 114 = 684$ eggs
d $6 × 1673 = 10\ 038$ eggs

3.6

a $48 ÷ 6 = 8$ boxes
b $42 ÷ 6 = 7$ boxes
c $258 ÷ 6 = 43$ boxes
d $15\ 732 ÷ 18 = 874$ boxes

3.7

Term	+	−	×	÷
add	•			
and	•			
difference		•		
divide				•
from		•		
goes into				•
how many more		•		
how many less		•		
less		•		
minus		•		
multiply			•	
plus	•			
product			•	
share				•
sum	•			
subtract		•		
take away		•		
times			•	

3.8
a (i) $27 + 18 - 9 = 36$ people left
   (ii)   ... $+ 12 - 21 = 27$ people left
   (iii) ... $+ 5 - 16 = 16$ people left
b (i) $205 - 21 = 184$ degrees
   (ii)   Average temperature rise per minute $= \frac{184}{8} = 23$ degrees.
c (i) $7 \times 15 = 105$ glasses of water in a week
   (ii) $31 \times 15 = 465$ glasses of water during the month of July
   (iii) $365 \times 15 = 5475$ glasses of water in a year.
d $7 \times 10 = 70$ cl

3.9
a $43 + 67 = 110$
b $162 + 77 = 239$
c $437 + 239 = 676$
d $259 + 952 = 1211$

3.10
a $\qquad$ 67 – 43 = 24
b $\qquad$ 162 – 77 = 85
c $\qquad$ 437 – 239 = 198
d $\qquad$ 952 – 259 = 693

3.11
a *Key sequence* $\qquad\qquad\qquad$ *Comment*

3 [+] 2 [=] [5] [–] 2 [=] [3]
4 [+] 3 [=] [7] [–] 3 [=] [4]
3 [×] 2 [=] [6] [÷] 2 [=] [3]
4 [×] 3 [=] [12] [÷] 3 [=] [4]
20 [÷] 5 [=] [4] [×] 5 [=] [20]

b 3 [+] 3 [+] 3 [+] 3 [=] [12]
How many lots of 3 did you add? [4]
4 [×] 3 [=] [12]

c 6 [–] 2 [–] 2 [–] 2 [=] [0]
6 [÷] 2 [=] [3]

3.12
1 No comments

2
Number	365	614	496	16042	1093461
Place value	ten	hundred	unit	thousand	ten

3
Temperature °C	10	4	21	–6	–10	0	3	–3
Three degrees less	7	1	18	–9	–13	–3	0	–6

4 The order doesn't matter when adding and multiplying but it does for subtracting and dividing. For example:

Adding $\qquad$ 2 + 3 = 3 + 2 = 5
Multiplying $\qquad$ 2 × 3 = 3 × 2 = 6
But
Subtracting $\qquad$ 2 – 3 = –1, whereas 3 – 2 = 1
Dividing $\qquad$ $2 \div 3 = \frac{2}{3}$, whereas $3 \div 2 = 1\frac{1}{2}$

5 (i) The square of an even number always gives an even number.
(ii) The square of an odd number always gives an odd number.
(iii) Not all odd numbers are prime (for example, 9, which is 3 × 3).
(iv) All prime numbers are odd with one exception, namely the number 2.

(v)

Number	prime	rectangular	odd	even	square
1			•		•
2	•			•	
3	•		•		
4		•		•	•
5	•		•		
6		•		•	
7	•		•		
8		•		•	
9		•	•		•
10		•		•	
11	•		•		
12		•		•	
13	•		•		
14		•		•	
15		•	•		
16		•		•	•
17	•		•		
18		•		•	
19	•		•		
20		•		•	

Note: By the way, the number 1 has not been marked as either a prime number or a rectangular number. It doesn't comfortably fit into either category, and can't be classified in this way.

04

fractions

In this chapter you will learn:
- how to picture a fraction
- about equivalent fractions
- how to calculate with fractions.

Decimals, now, and fractions. I don't know, I just didn't seem to grasp them ... I just found them boring. I couldn't concentrate on them at all. They weren't interesting enough for me. It wasn't attractive enough.

Sheila, a friend

Not the most promising start to a chapter on fractions perhaps! It certainly seems to be the case that, while most people know roughly what's going on when the four rules are applied to *whole* numbers, sums with fractions can bring the shutters down. The first thing you should realize is that fractions, decimals and percentages are all very similar. As you will see over the next three chapters, they are all slightly different ways of describing the same thing. What we call fractions – things like $\frac{1}{2}$, $\frac{3}{4}$ and so on – should really be called *common* fractions. In fact these sorts of fractions are not quite as 'common' as they used to be. Increasingly the more awkward common fractions, like $\frac{3}{8}$ and $\frac{5}{32}$, for example, are being replaced by *decimal* fractions. Decimal fractions look like 0.3, 0.125, and so on and are dealt with in Chapter 05. But first of all, let's find out what a fraction is and where it comes from.

What is a fraction?

Sabine and Sam are four. They have never heard of a fraction. I produced three squares of chocolate and said that they were to be shared between them. They took one square each. Now what about that third square? Well, you can be sure that they won't give it to me, or their favourite charity. Sabine and Sam may not have heard of a fraction, but they are quite capable of inventing one when the occasion arises. As will be explained below, fractions can be thought of as the 'broken bits' that lie between the whole numbers.

Fractions occur quite naturally in division (i.e. sharing) when the sum doesn't divide exactly. For example, sharing seven doughnuts amongst 3.

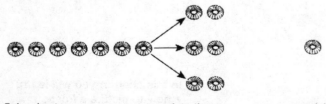

7 doughnuts
shared ...

... into three
lots of 2 ...

and 1 left over
(the remainder)

This can be written as follows

7 / 3 = 2 remainder 1.

But, as with the square of chocolate, we don't always want to leave the remainder 'unshared'. If the remainder of 1 is *also* shared out amongst the 3 (people) they each get an extra one third, as shown below.

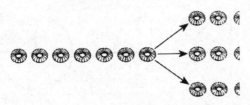

three lots of 'two and one third'

So, the more complete answer to this division sum is:

7 / 3 = 2⅓ i.e. 7 divided by 3 gives 2 and a third.

It is important to understand why fractions are written as they are. The fraction ⅓ really is another way of writing 1 divided by 3, or 1/3. So the top number in a fraction (called the *numerator*) is the number of things to be shared out. The bottom number (the *denominator*) tells you how many shares there will be.

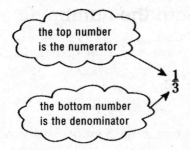

Exercise 4.1 will help you grasp this important idea.

EXERCISE 4.1 Sharing cheese

A box of processed cheese has six segments. Share two boxes equally among three people.

(i) How many segments will each person get?

(ii) What fraction of a box will each person receive?

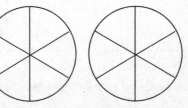

How to picture a fraction

The common fractions like a half, three quarters and two thirds are part of everyday language. You should find it helpful to have a mental picture of a fraction. The picture which is in my mind (and which is used in most schools when fractions are first introduced) is to imagine a whole as a complete cake. This can be cut into slices representing various fractions, like this:

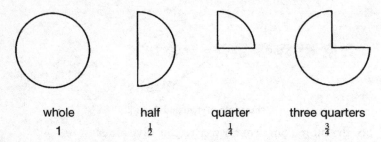

whole	half	quarter	three quarters
1	$\frac{1}{2}$	$\frac{1}{4}$	$\frac{3}{4}$

These sorts of picture are helpful as a way of understanding what a fraction is. But if you need to compare fractions or do calculations with them, then you need more than pictures. For example:

Is $\frac{2}{3}$ of a cake bigger than $\frac{3}{4}$ of it?

Fitting fractions into the number line

The next step is to understand how fractions fit into the sequence of numbers that you looked at in Chapter 02. For example:

$2\frac{1}{3}$ is one third of the way between 2 and 3

$3\frac{7}{10}$ is seven tenths of the way between 3 and 4

The diagram below shows how these fractions fit on the number line.

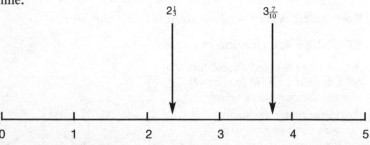

a number line showing fractions

Until fractions are introduced, numbers can be thought of as a set of points equally spaced on the line (i.e. the whole numbers). But as your picture of numbers expands, you can see that there are lots of other points between 1 and 2, between 2 and 3, and so on. How many are there? Are there any gaps at all on the number line when the fractions are added? These aren't questions with easy answers but you might care to think about them.

Finally, here is a reminder of how the cake diagram and the number line can help your mental picture of fractions.

Cake diagram → fractions are 'bits' of a whole

Number line → fractions fill in the gaps 'between whole numbers'

And now you're ready to add and subtract fractions. Well, nearly … Before that it would be useful to know what *equivalent* fractions are.

What are equivalent fractions?

What is the difference between sharing two cakes among four people or sharing one cake between two people? Well, since everybody ends up with half a cake, there is no difference in the share that each person gets. The first lot of people actually get $\frac{2}{4}$ of a cake but that would seem to be the same as $\frac{1}{2}$. So $\frac{2}{4}$ and $\frac{1}{2}$ are fractions which are the same; yet they are different – they have the same value but have different numerals top and bottom. The word used to describe this is *equivalence*.

We would say that $\frac{2}{4}$ and $\frac{1}{2}$ are *equivalent fractions*.

Exercise 4.2 (ii) will give you a chance to try to spot some more equivalent fractions. Try it now.

EXERCISE 4.2 FINDING EQUIVALENT FRACTIONS

(i) Mark with an arrow each of these numbers on the number line below:

$2\frac{3}{4}$, $\frac{1}{2}$, $4\frac{9}{10}$, $1\frac{1}{3}$

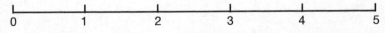

0	1	2	3	4	5

(ii) Find three fractions equivalent to each of the following (the first set has been done for you):

Fraction   Equivalent fractions

$\frac{3}{4}$, $\frac{6}{8}$, $\frac{9}{12}$, $\frac{12}{16}$

$\frac{1}{2}$

$\frac{9}{10}$

$\frac{1}{3}$

$\frac{6}{24}$

$\frac{10}{20}$

Adding and subtracting fractions

When adding fractions, it is helpful to think of the slices of cake. For example:

$\frac{1}{8} + \frac{3}{8} = ?$

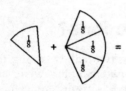

Here the slices are all the same size ($\frac{1}{8}$ each) so we just add them together, like this:

$\frac{1}{8} + \frac{3}{8} = \frac{4}{8}$

It is usual to write the answer in the form of the simplest equivalent fraction, so the answer, $\frac{4}{8}$, can be written as $\frac{1}{2}$. However, what happens when you have to add fractions like the following?

$\frac{2}{3} = \frac{1}{2} = ?$

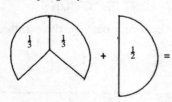

This time the slices of cake aren't the same size, so we can't just add them together. The way out of this problem is to cut both fractions until all the slices are the same size – like this:

Now, with all the slices equal to $\frac{1}{6}$, they *can* be added. The calculation looks like this:

$$\frac{2}{3} + \frac{1}{2}$$
$$= \frac{4}{6} + \frac{3}{6} \text{ Both fractions are changed to equivalent 'sixths'}$$
$$= \frac{7}{6}$$
$$= 1\frac{1}{6}$$

But why did I choose to subdivide each fraction into slices of $\frac{1}{6}$? The reason is that $\frac{1}{6}$ is the easiest fraction that a half and a third will break up into. I could have used slices of $\frac{1}{12}$ or $\frac{1}{18}$ but that would have been unnecessarily complicated. By the way, this process of breaking fractions up into smaller slices so that they can be added or subtracted is called 'finding the smallest common denominator'.

To summarize, finding the lowest common denominator means finding the smallest number which both the denominators will divide into. (Remember that the denominator is the bottom number in the fraction.) This number then becomes the new denominator. Thus in the example above, the lowest number that 3 and 2 both divide into is 6, so 6 is the new denominator. Now try Exercise 4.3 (the first one has been done for you).

EXERCISE 4.3 Adding and subtracting fractions

Complete the table below:

Calculation	Equivalent fractions	Answer
$\frac{2}{3} + \frac{1}{4}$	$\frac{8}{12} + \frac{3}{12}$	$\frac{11}{12}$
$\frac{1}{2} + \frac{3}{4}$		
$\frac{1}{3} + \frac{5}{6}$		
$\frac{4}{5} - \frac{1}{2}$		
$\frac{1}{5} + \frac{1}{4}$		
$\frac{1}{4} - \frac{1}{5}$		

Multiplying and dividing fractions

How often in your life have you had to multiply or divide two fractions outside a school mathematics lesson? I suspect that the answer is, for most people, never. I therefore don't intend to devote much space to this difficult and rather pointless exercise. However, it *is* useful to know a few basic facts – for example, a half of a half is a quarter, and a tenth of a tenth is one hundredth. There are also a few practical situations (like scaling the ingredients of a recipe, for example, when you want to produce a smaller or larger cake than the one in the recipe) where multiplication and division of very simple fractions may be helpful. This is probably easier to understand by looking at decimal fractions, so we shall return to this topic in Chapter 05.

Summary

- Fractions can be thought of as bits of whole numbers.
- A useful way of representing fractions is as slices of a cake.
- Equivalent fractions, like $\frac{1}{2}$ and $\frac{2}{4}$ have the same value and correspond to the same size of slice of the cake.
- Adding and subtracting fractions usually involves rewriting the fractions as equivalent fractions. This means finding a common denominator (the bottom number in the fraction), and adding the numerators (the top numbers in the new fractions).
- Multiplying and dividing fractions is easiest to understand when the fractions are written as decimal fractions (see Chapter 5).

Practice exercise

1 Share the following equally. (The first one has been done for you.)

a 11 cakes amongst 4 people. Each gets $2\frac{3}{4}$ cakes

b 17 cakes amongst 5 people. Each gets ☐ cakes

c 5 cakes amongst 6 people. Each gets ☐ cakes

d 20 cakes amongst 3 people. Each gets ☐ cakes

2 The number 1 consists of 3 thirds. How many thirds are there in the following numbers?

   a 2   **d** 12
   b 4   **e** $3\frac{1}{3}$
   c 10   **f** $7\frac{2}{3}$

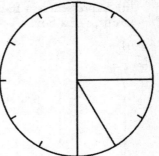

3 Write the appropriate fractions onto the slices of the clock.

Now add the fractions together.

Check that they add to 1.

4 The fraction $\frac{8}{10}$ can be written more simply as $\frac{4}{5}$. Write the following fractions in their simplest form. (One has been done for you.)

Fractions $\frac{8}{10}$ $\frac{4}{6}$ $\frac{5}{10}$ $\frac{12}{18}$ $\frac{6}{9}$ $\frac{4}{16}$ $\frac{8}{48}$ $\frac{9}{18}$ $\frac{2}{22}$

Simplest form $\frac{4}{5}$

5 (i) Change all the fractions below to twelfths. (The first one has been done for you.)

   (ii) Now rank them in order of size, putting a rank of 1 against the largest fraction and 6 against the smallest. (Again, one has been done for you.)

Fractions $\frac{2}{3}$ $\frac{3}{4}$ $\frac{2}{6}$ $\frac{7}{12}$ $\frac{5}{6}$ $\frac{1}{2}$

Fractions as twelfths $\frac{8}{12}$

Rank   3

6 A sum of £600 000 was left to be shared among three charities, as follows:
Charity A was to receive one quarter.
Charity B was to receive two thirds.
Charity C was to receive the rest.
Calculate:
   a the fraction of the sum that went to Charity C
   b the amount of money due to each charity.

Answers to exercises for Chapter 04

4.1 (i) Each person gets $\frac{6+6}{3} = \frac{12}{3} = 4$ segments.

(ii) Expressed as a fraction, each person gets $\frac{4}{6}$ or in other words, $\frac{2}{3}$ of a box.

4.2 (i)

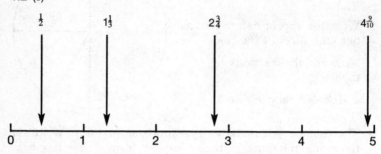

(ii) There are many possible answers here.

Fraction	Equivalent fractions		
$\frac{3}{4}$	$\frac{6}{8},$	$\frac{9}{12},$	$\frac{12}{16}$
$\frac{1}{2}$	$\frac{2}{4},$	$\frac{5}{10},$	$\frac{12}{24}$
$\frac{9}{10}$	$\frac{18}{20},$	$\frac{45}{50}$	$\frac{900}{1000}$
$\frac{1}{3}$	$\frac{2}{6},$	$\frac{20}{60},$	$\frac{100}{300}$
$\frac{6}{24}$	$\frac{1}{4},$	$\frac{2}{8},$	$\frac{12}{48}$
$\frac{10}{20}$	$\frac{1}{2},$	$\frac{9}{18},$	$\frac{12}{24}$

4.3

Calculation	Equivalent fractions	Answer
$\frac{2}{3} + \frac{1}{4}$	$\frac{8}{12} + \frac{3}{12}$	$\frac{11}{12}$
$\frac{1}{2} + \frac{3}{4}$	$\frac{2}{4} + \frac{3}{4}$	$\frac{5}{4} = 1\frac{1}{4}$
$\frac{1}{3} + \frac{5}{6}$	$\frac{2}{6} + \frac{5}{6}$	$\frac{7}{6} = 1\frac{1}{6}$
$\frac{4}{5} - \frac{1}{2}$	$\frac{8}{10} - \frac{5}{10}$	$\frac{3}{10}$
$\frac{1}{5} + \frac{1}{4}$	$\frac{4}{20} + \frac{5}{20}$	$\frac{9}{20}$
$\frac{1}{4} - \frac{1}{5}$	$\frac{5}{20} - \frac{4}{20}$	$\frac{1}{20}$

Answers to practice exercise

1 a 11 cakes amongst 4 people. Each gets $\boxed{2\frac{3}{4}}$ cakes

b 17 cakes amongst 5 people. Each gets $\boxed{3\frac{2}{5}}$ cakes

  c 5 cakes amongst 6 people. Each gets $\boxed{\frac{5}{6}}$ cakes

  d 20 cakes amongst 3 people. Each gets $\boxed{6\frac{2}{3}}$ cakes

2 a $2 = \frac{6}{3}$ **d** $12 = \frac{36}{3}$
  b $4 = \frac{12}{3}$ **e** $3\frac{1}{3} = \frac{10}{3}$
  c $10 = \frac{30}{3}$ **f** $7\frac{2}{3} = \frac{23}{3}$

3 The fractions are $\frac{1}{4}$, $\frac{1}{6}$, $\frac{1}{12}$ and $\frac{1}{2}$.
  These can be rewritten in twelfths and added, as follows.
    $\frac{3}{12} + \frac{2}{12} + \frac{1}{12} + \frac{6}{12} = \frac{3+2+1+6}{12} = \frac{12}{12} = 1$

4

Fractions	$\frac{8}{10}$	$\frac{4}{6}$	$\frac{5}{10}$	$\frac{12}{18}$	$\frac{6}{9}$	$\frac{4}{16}$	$\frac{8}{48}$	$\frac{9}{18}$	$\frac{2}{22}$
Simplest form	$\frac{4}{5}$	$\frac{2}{3}$	$\frac{1}{2}$	$\frac{2}{3}$	$\frac{2}{3}$	$\frac{1}{4}$	$\frac{1}{6}$	$\frac{1}{2}$	$\frac{1}{11}$

5

Fractions	$\frac{2}{3}$	$\frac{3}{4}$	$\frac{2}{6}$	$\frac{7}{12}$	$\frac{5}{6}$	$\frac{1}{2}$
Fractions as twelfths	$\frac{8}{12}$	$\frac{9}{12}$	$\frac{4}{12}$	$\frac{7}{12}$	$\frac{10}{12}$	$\frac{6}{12}$
Rank	3	2	6	4	1	5

6 a Charity C was to receive $1 - (\frac{1}{4} + \frac{2}{3}) = 1 - \frac{11}{12} = \frac{1}{12}$
  b Charity A was to receive $\frac{1}{4} \times £600\ 000 = £150\ 000$
     Charity B was to receive $\frac{2}{3} \times £600\ 000 = £400\ 000$
     Charity C was to receive $\frac{1}{12} \times £600\ 000 = £50\ 000$
(As a quick check, these amounts of money should add to
£600 000. £150 000 + £400 000 + £50 000 = £600 000.)

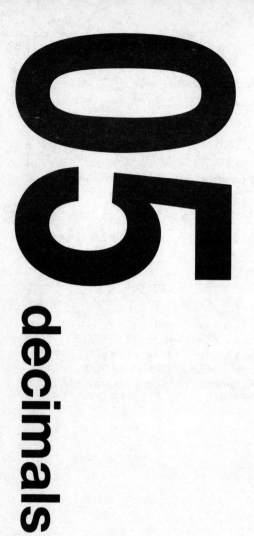

05

decimals

In this chapter you will learn:
- about the 'ten-ness' of numbers
- why we use a decimal point and where we put it
- about the connection between fractions and decimals
- how to calculate with decimals.

Five fingers on each hand (well, four fingers and one thumb) seems to be a reasonable number to possess. Any fewer and we wouldn't be able to play the 'Moonlight Sonata' with the same panache; any more and we'd have a bit of a struggle putting on a pair of gloves. It may not surprise you that the link between the number of our *ten*tacles and decimals is, well, more than *ten*uous. What I'm really saying, then, is that the reason our number system is based on the number ten is because humans have counted on their ten fingers for thousands of years. 'Decimals' (from the Latin *deci* meaning ten) is really a way of describing the ten-ness of our counting system. However, it usually refers to decimal fractions. And there is no shortage of those around us. Just listen to sports commentators, for example:

> ... the winning time of 10.84 seconds smashes the world record by two hundredths of a second.
> ... a long jump of 8 metres, point 21.
> ... the winning scores for the pairs ice skating are as follows: 5.9, 5.8, 5.9 ...

Decimal points appear whether we are talking about money (£8.14) or measurement (2.31 metres) and will appear on a calculator display at the touch of a button. Now find yourself a calculator and you can use it to investigate exactly what a decimal fraction is.

Decimal fractions

What is a decimal fraction?

A decimal fraction is simply another way of writing a common fraction. In this section you can use your calculator to discover how fractions and decimals are connected.

EXERCISE 5.1   Deriving decimals from fractions

For each box of questions below:
a write down the answers in fractions
b use your calculator to find the answers in decimals
c complete the blank in the 'Conclusion' box.

The first one has been started for you.

Key sequence	Fraction	Decimal	Conclusion
1 ÷ 2 =	$\frac{1}{2}$	0.5	
2 ÷ 4 =			
5 ÷ 10 =	$\frac{5}{10}$		
50 ÷ 100 =			The decimal for $\frac{1}{2}$ is $\boxed{0.5}$
1 ÷ 4 =			
2 ÷ 8 =			
5 ÷ 20 =			
25 ÷ 100 =			The decimal for $\frac{1}{4}$ is $\boxed{}$
3 ÷ 4 =			
6 ÷ 8 =			
75 ÷ 100 =			The decimal for $\frac{3}{4}$ is $\boxed{}$
1 ÷ 10 =			
10 ÷ 100 =			The decimal for $\frac{1}{10}$ is $\boxed{}$

As can be discovered from the key sequences above, converting from fractions to decimal fractions is very straightforward using a calculator. For example, the fraction $\frac{5}{8}$ can be converted to a decimal fraction by dividing 5 by 8, i.e. by pressing the following key sequence.

5 ÷ 8 =

This produces the result 0.625.

In other words, the fraction $\frac{5}{8}$ has the same value as the decimal fraction 0.625. Exercise 5.2 will give you practice at converting from fractions to decimal fractions.

EXERCISE 5.2 Converting from fractions to decimal fractions

Now use your calculator to find the decimal values of the fractions in the table below.

Table 1

Fraction	$\frac{1}{2}$	$\frac{1}{4}$	$\frac{3}{4}$	$\frac{1}{10}$	$\frac{1}{5}$	$\frac{2}{5}$	$\frac{3}{10}$	$\frac{9}{10}$	$\frac{1}{20}$	$\frac{1}{8}$	$\frac{1}{3}$
Decimal											

Picturing decimal fractions

You may remember from the previous chapter that fractions could be helpfully represented using slices of a cake. Since there is such a close link between fractions and decimal fractions, it follows that the same helpful pictures apply to decimal fractions. Here are the 'cakes' from Chapter 04, but this time with the corresponding decimal fractions added.

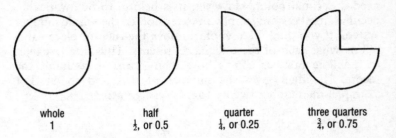

whole	half	quarter	three quarters
1	$\frac{1}{2}$, or 0.5	$\frac{1}{4}$, or 0.25	$\frac{3}{4}$, or 0.75

Let's now turn to the way we represent decimal fractions on a number line. Again, since fractions and decimals are really very similar, it is not surprising that they can both be represented in the same way. For example, the fraction $\frac{3}{4}$ and the decimal 0.75 share the same position on the number line. Thus:

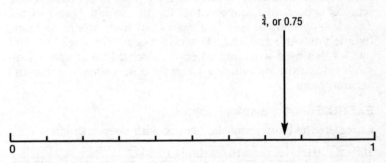

Having made a connection between fractions and decimals, the next exercise (called 'Guess and press') gets you working just with decimals. The idea is to write down your *guess* as to what the answer will be for each calculation. Then you *press* the key sequence on the calculator and see if you are right. The aim of this exercise is to help you see the connection between decimals and whole numbers.

EXERCISE 5.3 Guess and press

Calculation	*Guess*	*Press*
0.5 $\boxed{+}$ 0.5 $\boxed{=}$	1	1
0.5 $\boxed{\times}$ 2 $\boxed{=}$		
0.25 $\boxed{\times}$ 4 $\boxed{=}$		
0.5 $\boxed{\times}$ 10 $\boxed{=}$		
4 $\boxed{\div}$ 10 $\boxed{=}$		
0.1 $\boxed{+}$ 0.1 $\boxed{=}$		
0.1 $\boxed{\times}$ 10 $\boxed{=}$		

What is the point of the decimal point?

If the world contained only whole numbers, we would never need a decimal point. However, it is helpful to be aware that decimal numbers are simply an extension of the whole number system. If you think of a whole number, the rules of place value tell us what each of these digits represents. Thus, the last digit of a whole number shows how many *units* it contains, the second last digit gives the number of *tens* and so on. For example, the number twenty-four is written as:

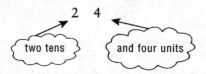

2 4

two tens   and four units

However, when we start to use numbers which include bits of a whole (i.e. with decimals) some other 'places' are needed. These represent the tenths, hundredths, the thousandths, and so on. *The decimal point is simply a marker* to show where the units (whole numbers) end and the tenths begin. You'll get a better idea of this by discovering when the decimal point appears on your calculator. As you do Exercise 5.4, watch out for the decimal point …

EXERCISE 5.4   Blank checks

Complete the blanks and then check with your calculator.

100 ÷ 10 = ☐ ÷ 10 = ☐ ÷ 10 = ☐ ÷ 10 = ☐
400 ÷ 10 = ☐ ÷ 10 = ☐ ÷ 10 = ☐ ÷ 10 = ☐
25 ÷ 10 = ☐ ÷ 10 = ☐ ÷ 10 = ☐ ÷ 10 = ☐
1   ÷ 100 = ☐
4   ÷ 100 = ☐
7   ÷ 1000 = ☐

You may have got a picture of the decimal point jumping one place to the left every time you divide by 10. Actually, this is a slightly misleading picture. Most calculators work on a principle of a 'floating decimal point'. This means that the decimal moves across the screen to keep its position between the units and the tenths digit.

The key point to remember is that the decimal point is nothing more than a mark separating the units from the tenths. Below I've written out the more complete set of 'place values' extending beyond hundreds, tens and units into decimals.

Hundreds	Tens	Units	Decimal point	Tenths	Hundredths	Thousandths
(100)	(10)	(1)	•	(0.1)	(0.01)	(0.001)

Using the four rules with decimals

If you aren't sure how to use the four rules of +, −, × and ÷ with decimal numbers, why don't you experiment with your calculator? You will quickly discover that the four rules work in exactly the same way for decimals as for whole numbers, and for that reason addition and subtraction of decimals are not spelt out here as a separate topic.

Back in Chapter 04, I suggested that you could learn about multiplying and dividing fractions by experimenting with your calculator. Exercise 5.5 is designed to help you do just that.

EXERCISE 5.5   Multiplying with decimal fractions

The first column in this table gives you three multiplication calculations involving fractions. For each calculation:

a change the fractions to decimals (Column 2)
b use your calculator to multiply the decimals (Column 3)
c change the decimal answer back to a fraction (Column 4)

Calculation in fractions	Calculation in decimal form	Decimal answer (use a calculator)	Fraction answer
$\frac{1}{2} \times \frac{1}{2}$	0.5×0.5	0.25	$\frac{1}{4}$
$\frac{1}{2} \times \frac{1}{5}$			
$\frac{3}{5} \times \frac{1}{2}$			

Now look at columns (1) and (4) and see if you can spot the rule for multiplying fractions. Think about this for a while before reading on.

Rule for multiplying fractions

To multiply two fractions, say $\frac{3}{5}$ and $\frac{1}{2}$, multiply the numerators (3 × 1, giving 3), then multiply the denominators (5 × 2, giving 10). The answer in this case is $\frac{3}{10}$: i.e. $\frac{3}{5} \times \frac{1}{2} = \frac{3 \times 1}{5 \times 2} = \frac{3}{10}$

EXAMPLE 1

Look at the following multiplication:

$\frac{3}{4} \times \frac{2}{5}$

As before, these two fractions can be converted into decimal form, so the calculation can be rewritten as follows.

0.75×0.4

Pressing 0.75 $\boxed{\times}$ 0.4 $\boxed{=}$ on the calculator gives an answer of 0.3, or $\frac{3}{10}$.

By way of a check, we could apply the rule on the previous page. We multiply the two numerators and then the two denominators, as follows.

$\frac{3}{4} \times \frac{2}{5} = \frac{3 \times 2}{4 \times 5} = \frac{6}{20}$

This can be simplified to $\frac{3}{10}$ (remember from Chapter 04 that $\frac{6}{20}$ and $\frac{3}{10}$ are equivalent fractions).

So, it doesn't matter whether multiplication is done in fraction form or decimal form; the result is the same either way.

EXAMPLE 2

Look at this multiplication:

$2\frac{1}{4} \times 4\frac{3}{5}$

Again, these two fractions can be converted into decimal form, so the calculation can be rewritten as follows.

2.25×4.6

Pressing 2.25 $\boxed{\times}$ 4.6 $\boxed{=}$ on the calculator gives an answer of 10.35.

As before, we can check this against the method of multiplying fractions described in the box. However, the number $2\frac{1}{4}$ must be rewritten as $\frac{9}{4}$ and $4\frac{3}{5}$ must be rewritten as $\frac{23}{5}$.

$\frac{9}{4} \times \frac{23}{5} = \frac{9 \times 23}{4 \times 5} = \frac{207}{20}$

Finally, just to check that the two methods produce the same result, this fraction can be converted to decimal form by dividing the numerator by the denominator, thus:

Pressing 207 $\boxed{\div}$ 20 $\boxed{=}$ on the calculator confirms the previous answer of 10.35.

Dividing fractions

Dividing fractions is a more painful and less useful skill than multiplying them and I don't propose to waste much time on it here. This topic is one of several 'casualties' of the calculator age which is no longer relevant and which, in my view, is simply not worth learning. If you *are* in a situation where you need to divide fractions, a good strategy is to convert the fractions to decimals and perform the division on your calculator. Here are two examples.

EXAMPLE 1

$$\tfrac{3}{4} \div \tfrac{2}{5}$$

Rewriting as decimal fractions, this gives:

$$0.75 \div 0.4$$

Using the calculator, this gives an answer of 1.875.

A rule of thumb which used to be taught for dividing fractions is to turn the fraction you are dividing by upside down and proceed as for multiplication. So, in this case we get:

$$\tfrac{3}{4} \div \tfrac{2}{5} = \tfrac{3}{4} \times \tfrac{5}{2} = \tfrac{15}{8}$$

If you check on your calculator (by pressing 15 $\boxed{\div}$ 8 $\boxed{=}$), you will see that the fraction $\tfrac{15}{8}$ has the same value as the earlier answer of 1.875.

EXAMPLE 2

Look at this division:

$$4\tfrac{1}{5} \div 1\tfrac{1}{2}$$

Again, these two fractions can be converted into decimal form, so the calculation can be rewritten as follows:

$$4.2 \div 1.5$$

Pressing 4.2 $\boxed{\div}$ 1.5 $\boxed{=}$ on the calculator gives an answer of 2.8.

As before, we can check this against the method of dividing fractions described above. However, the number $4\tfrac{1}{5}$ must be rewritten as $\tfrac{21}{5}$ and $1\tfrac{1}{2}$ must be rewritten as $\tfrac{3}{2}$.

$$\tfrac{21}{5} \div \tfrac{3}{2} = \tfrac{21}{5} \times \tfrac{2}{3} = \tfrac{21 \times 2}{5 \times 3} = \tfrac{42}{15}$$

which can be simplified to $\tfrac{14}{5}$.

Finally, just to check that the two methods produce the same result, this fraction can be converted to decimal form by dividing the numerator by the denominator, giving the same answer as before.

$$14 \div 5 = 2.8.$$

An overview of decimals

As your confidence with decimals grows, you will come to appreciate how decimal numbers are a natural extension of our whole number system. What this means in practice is being able to understand place value. So just as you add up to 10 units and then swap them for one ten, so you add up to ten hundredths and swap them for one tenth. This is illustrated in the two addition sums below:

1	0.01
6	0.06
3	0.03
10	0.10

ten units are written as 1 in the tens column

ten hundredths are written as 1 in the tenths column

(Note that this result, 0.10, will be shown simply as 0.1 on the calculator display)

One obvious property of whole numbers is that the more digits a number has, the bigger it is. Unfortunately this is *not* true for decimal numbers. For example, the number 5.831659 is actually smaller than, say, 7.2. Don't be unduly impressed by a long string of digits. What matters is the position of the decimal point. You need to see beyond this string of digits and get a sense of how big the number actually is. For example, it is more useful to know that 5.831659 is between 5 and 6 (or just less than 6) than to quote it to six decimal places.

Sensibly used, calculators are an excellent means of seeing beyond the digits of a number. For example, earlier in the chapter, in Exercise 5.4, you were asked to perform repeated division by 10 and then observe what happened to the decimal point of the answer. This is an exercise which you can do with any starting number of your own choice, and the repeated division by 10 can be more efficiently done by using the calculator's constant facility.

Exercise 5.6 contains some calculator activities which should help you to become more confident with decimals.

EXERCISE 5.6 Using the constant to investigate decimals

a Set your calculator's constant to divide by 10. Next enter a large number into the calculator display and then repeatedly press ☐=☐. At first, watch what happens. Later, try to predict what will happen.

b Set the constant to multiply by 10. Then enter a small decimal fraction and repeatedly press ☐=☐. Try to make sense of what is going on and then try to predict what will happen next.

c Set the constant to add 0.1 and repeatedly press ☐=☐. Without pressing the 'Clear' key, enter a large decimal number and keep pressing ☐=☐.

d Set the constant to add 0.01 and repeat what you have just done in part c.

e Repeat parts c and d but with the constant set for subtraction in each case.

f With a friend, play the game 'Guess the number', the rules of which are explained at the end of the chapter.

Practical situations involving decimals abound, the most obvious example being money. Thus £3.46 represents 3 whole pounds, 4 tenths of a pound (i.e. 4 ten-pences) and 6 hundredths of a pound (i.e. 6 pence).

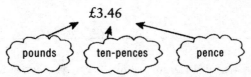

However, the money representation of decimals can be confusing. We *say* £3.46 as 'three pounds forty-six', rather than 'three point four six', which is the more correct decimal form. This latter version emphasizes the decimal place value of each digit. Otherwise you can get into trouble when dealing with sums of money like one pound and nine pence, which is often mistakenly written as £1.9, rather than £1.09.

We grow up with decimals and metric units like metres, centimetres, kilograms, millilitres, and so on all around us. However, we still have feet and inches, pounds and ounces, and these units, known as imperial units, are the ones that many adults still feel happy with. These units are explained in some detail in Chapter 07.

The main advantage of metric units is that they are based entirely on tens, hundreds and thousands; for example, there are 100 centimetres in a metre, 1000 metres in a kilometre, 1000 millilitres in a litre, and so on. Contrast this with the old-fashioned 14 pounds in a stone, 12 inches in a foot, 1760 yards in a mile, and so on – really a complete shambles!

Practice exercise

1 a Mark the numbers 0.35 and 0.4 on the number line below.

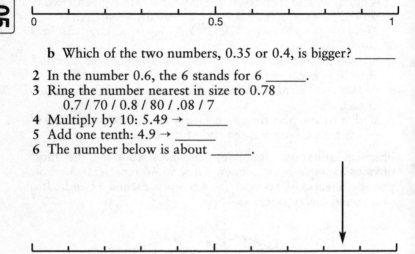

 b Which of the two numbers, 0.35 or 0.4, is bigger? _____

2 In the number 0.6, the 6 stands for 6 _____.
3 Ring the number nearest in size to 0.78
  0.7 / 70 / 0.8 / 80 / .08 / 7
4 Multiply by 10: 5.49 → _____
5 Add one tenth: 4.9 → _____
6 The number below is about _____.

7 How many different numbers can you write down between 0.26 and 0.27?
8 Which of these numbers is larger, 24.91257 or 83? _____

Summary

This chapter should have helped you to make the link between fractions and decimals. I hope that, after reading it, you now have a clearer sense of how decimal fractions (i.e. numbers like 0.56, 45.03 and so on) fit into the way the number system is organized. While digits to the left of the decimal point represent the number of units, tens, hundreds, and so on, the digits to the right are the tenths, hundredths, thousandths, and so on.

If you want to explore numbers, the calculator is an excellent place to start. A variety of calculator activities were suggested which should all contribute to your understanding of, and confidence with, decimals.

Finally, here is a checklist of the sort of things you should aim to know about decimals. *Note*: You will have the opportunity of using decimals again when we look at units of measure in Chapter 07.

Decimals checklist

You should have the ability to:
- know that the 4 in the number 6.143 refers to four hundredths
- mark decimal numbers on the number line
- arrange decimal numbers in order from smallest to biggest
- multiply and divide decimal numbers by 10, 100 and 1000
- know that 3.45 is half way between 3.4 and 3.5
- know that 0.25 means $\frac{1}{4}$ and 3.75 means $3\frac{3}{4}$
- handle units (metres, pounds (£), kilograms) in practical situations
- know roughly what answer to expect in a calculation involving decimals.

Answers to exercises for Chapter 05

Your calculator should have provided you with most of the answers to these exercises. However, here are some of the main points.

5.1 Key sequence	Fraction	Decimal	Conclusion
1 ÷ 2 =	$\frac{1}{2}$	0.5	
2 ÷ 4 =	$\frac{2}{4}$	0.5	
5 ÷ 10 =	$\frac{5}{10}$	0.5	
50 ÷ 100 =	$\frac{50}{100}$	0.5	The decimal for $\frac{1}{2}$ is $\boxed{0.5}$
1 ÷ 4 =	$\frac{1}{4}$	0.25	
2 ÷ 8 =	$\frac{2}{8}$	0.25	
5 ÷ 20 =	$\frac{5}{20}$	0.25	

25 ÷ 100 = $\frac{25}{100}$ 0.25   The decimal for $\frac{1}{4}$ is $\boxed{0.25}$

3 ÷ 4 = $\frac{3}{4}$ 0.75

6 ÷ 8 = $\frac{6}{8}$ 0.75

75 ÷ 100 = $\frac{75}{100}$ 0.75   The decimal for $\frac{3}{4}$ is $\boxed{0.75}$

1 ÷ 10 = $\frac{1}{10}$ 0.1

10 ÷ 100 = $\frac{10}{100}$ 0.1   The decimal for $\frac{1}{10}$ is $\boxed{0.1}$

5.2
Fraction	$\frac{1}{2}$	$\frac{1}{4}$	$\frac{3}{4}$	$\frac{1}{10}$	$\frac{1}{5}$	$\frac{2}{5}$	$\frac{3}{10}$	$\frac{9}{10}$	$\frac{1}{20}$	$\frac{1}{8}$	$\frac{1}{3}$
Decimal	0.5	0.25	0.75	0.1	0.2	0.4	0.3	0.9	0.05	0.125	0.33

5.3
Calculation	*Press*
0.5 + 0.5 =	1.
0.5 × 2 =	1.
0.25 × 4 =	1.
0.5 × 10 =	5.
4 ÷ 10 =	0.4
0.1 + 0.1 =	0.2
0.1 × 10 =	1.

5.4 100 ÷ 10 = 10 ÷ 10 = 1 ÷ 10 = 0.1 ÷ 10 = 0.01
400 ÷ 10 = 40 ÷ 10 = 4 ÷ 10 = 0.4 ÷ 10 = 0.04
25 ÷ 10 = 2.5 ÷ 10 = 0.25 ÷ 10 = 0.025 ÷ 10 =
0.0025
1 ÷ 100 = 0.01
4 ÷ 100 = 0.04
7 ÷ 1000 = 0.007

5.5
Calculation in fractions	Calculation in decimal form	Decimal answer (use a calculator)	Fraction answer
$\frac{1}{2} \times \frac{1}{2}$	0.5 × 0.5	0.25	$\frac{1}{4}$
$\frac{1}{2} \times \frac{1}{5}$	0.5 × 0.2	0.1	$\frac{1}{10}$
$\frac{3}{5} \times \frac{1}{2}$	0.6 × 0.5	0.3	$\frac{3}{10}$

5.6 No comments

Answers to practice exercise

1 a

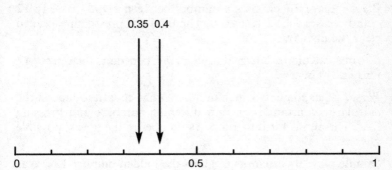

0.35 0.4

0   0.5   1

b 0.4 is bigger than 0.35
2 In the number 0.6, the 6 stands for 6 tenths.
3 Ring the number nearest in size to 0.78
0.7 / 70 /(0.8)/ 80 / .08 / 7
4 Multiply by 10: 5.49 → 54.9
5 Add one tenth: 4.9 → 5.0
6

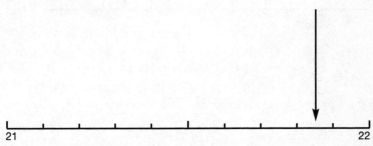

21   22

This number is about 21.85

7 There are infinitely many numbers between 0.26 and 0.27.
For example, I could write out the thousandths: 0.261, 0.262,
0.263, and so on up to 0.269. There are nine of these. But
between, say, 0.262 and 0.263 I could write out nine further
numbers, each expressed as a ten thousandth; 0.2621,
0.2622, 0.2623, and so on. Then I can write out numbers in
hundred thousandths, millionths, and so on. This process can
continue indefinitely or until I fall over with exhaustion.
8 Although 24.91257 contains more digits than 83, its value is
only about 25, so 83 is larger.

Guess the number

A game for two players, based on the calculator constant.

Player A secretly chooses a number between 1 and 20 – say 12 – and presses 1 ⌷÷⌷ 12 ⌷=⌷ 0. The final 0 is pressed in order to clear the display.

(If your calculator has a 'double press' constant, then press 12 ⌷÷⌷ ⌷÷⌷ 0 instead.)

Player B has to guess which number A has chosen to hide in the calculator constant by trying different numbers and pressing ⌷=⌷. The aim is for B to guess A's number in the fewest possible guesses.

Sample play: B's attempts to guess the hidden number 12 are as follows.

B presses	Display	Comments
16 ⌷=⌷	1.3333333	16 is too big
15 ⌷=⌷	1.25	15 is too big
9 ⌷=⌷	0.75	9 is too small
12 ⌷=⌷	1.	12 is the hidden number

06

percentages

In this chapter you will learn:
- why percentages are important
- about the connection between percentages, fractions and decimals
- how to do percentage calculations
- about common difficulties that people experience with percentages.

There are roughly 20 million telephones in Russia and only about one million in Ireland. So it would seem that people in Russia are better off in respect of access to telephones than the Irish.

The facts are right but the conclusion is wrong when you realize that the population of Russia is about eighty times that of Ireland. In fact about 75 per cent of households in Ireland (that is, 75 out of every 100 households) have a telephone, whereas only about 15 per cent of Russian households have one.

Failing to compare like with like can result in quite incorrect conclusions, as this example has shown. Percentages are a useful device for making fair comparisons. Unfortunately, many people find percentages difficult. Government reports and educational researchers have confirmed that amongst adults there is a widespread inability to understand percentages. Yet this is despite the fact that you can't pick up a newspaper or watch TV without coming across the word percentage over and over. I opened a daily newspaper at random and quickly picked out the following two examples.

Read the cuttings now and try to make sense of how the word percentage is being used. You will get another chance to read them at the end of the chapter.

Labour attacks mortgage insurance plan

Michael Simmons, Community Affairs Correspondent

Labour launched an assault on the Government's mortgage insurance scheme yesterday, accusing ministers of betraying homeowners with misleading promises. According to Labour research, housing costs will soon be 25 per cent higher than they were a year ago.

Gordon Brown, shadow chancellor, told a news conference that the combination of higher interest rates, cuts in mortgage tax relief and the "threat" of mortgage insurance meant that typical housing costs, which were £3,200 a year last April, would pass the £4,000 mark in October.

Italy's long dolce vita

RESEARCHERS are trying to pinpoint the reasons for the longevity among residents in Campodimele, a mountaintop village about 70 miles southeast of Rome. The studies suggest the hamlet is a near-ideal union of the classic Mediterranean diet and healthy bucolic living.

More than 10 per cent of the village's 900 residents are between 75 and 99 years old, the age of the oldest resident.

"Campodimele is not unique, but I believe it's about as close as you can get to the perfect environment for a long life," Dr Alessandro Menotti said.

Source: *Guardian* 2/2/95

The reason for the confusion that most people have with percentages is, I think, quite simple. Many adults and most children don't really understand what a percentage is.

What is a percentage?

The first thing you should realize about a percentage is that it is very similar to a decimal and a fraction. Like them, a percentage is used to describe a 'bit' of a number. But really it is nothing more than a particular sort of fraction. Think back to Chapter 04 on fractions, which explained how two or more fractions could be *equivalent*. Here are four fractions which are equivalent:

$$\frac{1}{2}, \quad \frac{2}{4}, \quad \frac{5}{10}, \quad \frac{50}{100}$$

These fractions are equivalent because they share the same value of a half.

Now look at the last of these four fractions, $\frac{50}{100}$. You might read it as 'fifty out of a hundred'. A shorthand way of saying this uses the Latin word *per centum* meaning 'out of every hundred'.

So, $\frac{50}{100}$ is the same thing as 'fifty per cent'.

Well, if a half (i.e. $\frac{50}{100}$) is the same as 50 per cent, what do you think a quarter becomes as a percentage?

The answer is 25 because $\frac{1}{4} = \frac{25}{100}$, or 25 per cent.

The symbol for 'per cent' is %.

So 25% is really another way of writing $\frac{25}{100}$ or 25 per cent.

Changing a fraction to a percentage

By now you might have worked out for yourself how to change a fraction to a percentage. If not, you can read the method which I've summarized in two simple steps below. Let us take the example of converting the fraction $\frac{4}{5}$ to a percentage.

Step 1   Change the fraction into its decimal form $\frac{4}{5} \rightarrow 0.8$
          (If you find this hard to do in your head, press 4 $\boxed{\div}$
          5 $\boxed{=}$ on your calculator.)

Step 2   Express the answer in hundredths.
0.8 is the same as 0.80, or 80 hundredths.
So, $\frac{4}{5}$ converts to 80%.

You will probably want to practise this, so try Exercise 6.1 now.

EXERCISE 6.1   Changing fractions to percentages

Fill in the blanks in the table below. The first one has been done for you.

Fraction	Decimal fraction	Percentage
$\frac{1}{2}$	0.5	50%
$\frac{3}{4}$		
$\frac{7}{10}$		
$\frac{1}{5}$		
$\frac{1}{20}$		
$\frac{3}{5}$		
$\frac{3}{8}$		

Like fractions and decimals, percentages can be represented on the number line. In the percentage number line below, 100% corresponds to the number 1, 200% to 2 and so on.

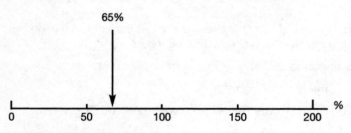

percentage number line

In practice, percentages are rarely represented in the form of a number line but I've included it to stress the similarity with fractions and decimals.

Why bother with percentages?

The main advantage of percentages is that they are much easier to compare than fractions. For example, which do you think is bigger, $\frac{3}{4}$ or $\frac{7}{10}$? Written like this you can't really say, because the slices of the whole 'cake' (quarters and tenths, respectively) are not the same size. In order to make a proper comparison, the fractions need to be broken down to the same size of slice, and hundredths are very convenient. So here goes …

Fraction	Hundredths	Percentage
$\frac{3}{4}$	$\frac{75}{100}$	75%
$\frac{7}{10}$	$\frac{70}{100}$	70%

Clearly 75% is bigger than 70%, so we can now conclude that $\frac{3}{4}$ is bigger than $\frac{7}{10}$.

If you look at a practical example you will get a better idea of how useful percentages are.

Which of the following would represent the bigger price rise?
a Bread to go up by 6p per loaf
b A refrigerator to go up by £5

In one sense the answer could be **b**, because £5 is more than 6p. But, since most people buy many more loaves of bread than they do fridges, we would probably be more concerned if bread went up by 6p per loaf. The only fair way to compare these price rises is to acknowledge that 6p is a lot compared with the price of a loaf of bread, whereas £5 may not be so much compared with the price of a refrigerator. Using percentages allows us to make comparisons, taking account of the prices of each item.

So, if we convert these price rises to percentage price rises, a very different picture emerges. In Exercise 6.2 you are asked to have a go at calculating these two percentage increases. Don't worry if you can't do it straightaway, as the method is explained below.

EXERCISE 6.2   Calculating percentage increases

Complete the table below. (I've taken the original price of bread to be 60p per loaf and that of the refrigerator to be £100.)

	Original price (£)	Price rise (£)	Percentage price rise
Bread	0.60	0.06	
Fridge	100.00	5.00	

Solution
The calculation of the percentage price increases is illustrated in the table over.

	Original price (£)	Price rise (£)	Percentage price rise
Bread	0.60	0.06	$\frac{0.06}{0.60} \times 100 = 10\%$
Fridge	100.00	5.00	$\frac{5}{100} \times 100 = 5\%$

In summary, then, percentage price increases (or decreases) are calculated as follows.

$$\text{Percentage price increase} = \frac{\text{Price increase}}{\text{Original price}} \times 100.$$

Perhaps you were able to confirm my calculation that bread went up in price by 10 per cent whereas the fridge went up by only 5 per cent in price.

There is, of course, another reason that this increase in the price of bread will cause more concern than that of the refrigerator. It is that we tend to buy bread every week, so this price rise is affecting our shopping bill every week. Fridges, on the other hand, are a very rare purchase and even a £5 price rise will simply not affect most people most of the time.

Let's now look in more detail at how to calculate percentage increases and reductions.

Calculating percentage increases and reductions

Where the percentages convert to very simple fractions (e.g. 100 per cent, 50 per cent, 25 per cent or 10 per cent), it should be possible to do the calculation in your head. However, for anything more complicated, I would always use a calculator to calculate percentage changes. Here, first, are two examples which could probably be done in your head.

EXAMPLE 1 50% reduction

SOCKS

Improve your *socks* life!

£~~2.50~~ a pair

Now 50% off!!

What is the sale price of a pair of these socks?

Solution

Since 50% is $\frac{1}{2}$, there is a reduction of half of £2.50.

This reduction is $\frac{£2.50}{2}$ = £1.25.

So the new price is £2.50 – £1.25 = £1.25.

EXAMPLE 2 5% increase

In 2001, the average price of a new house in a particular town in the Midlands was £104 500.

Over the next year, prices of new houses in the town increased by about 5%.

Estimate the average price of a new house a year later.

Solution

Since 5% is the same as $\frac{1}{20}$, we know that the prices rose by $\frac{1}{20}$ over this period.

So, price rise = $\frac{104500}{20}$ = £5225

Adding to the original price, we get:

Estimate of the average price of a new house a year later = £104 500 + £5 225 = £109 725.

So much for calculating simple percentage increases using pencil and paper only. Unfortunately, most percentage calculations are more complicated than this and require a calculator. The method for calculating percentage price changes is explained in the next two examples.

EXAMPLE 3   6% increase

Following a budget announcement on petrol tax, garages increased all their pump prices by 6%.

> Current petrol prices
> Unleaded premium, 82.1 per litre.
> All prices to go up by 6% at
> midnight tonight.

What is the new price of unleaded premium petrol at this garage?

Solution

An increase of 6% means an increase of six hundredths, or, in other words an increase of 0.06 of the original price. A possible way of proceeding here is to perform the calculation in two stages. First, find the price increase and then add it on to the original price. As you will see shortly, there is a quicker, one-staged method, but you will find this method easier to follow if you first work through the two stages explained below.

Stage 1:   Finding the price increase
The price increase = $0.06 \times 82.1p = 4.926p$
Petrol prices are usually quoted to one decimal place, so this price increase would be *rounded* to the nearest tenth of a penny, i.e. 4.9p.

Stage 2:   Adding on the price increase
The new price = 82.1p + 4.9p = 87.0p
(Notice that the price is written as 87.0p, rather than 87p in order to stress that the price has been stated accurate to one decimal place.)

As was suggested above, if all you want to find is the new price, this two-staged method is unnecessarily complex. The whole process can be reduced to a single stage, by multiplying the old price by 1.06.

It may not be obvious to you where the 1.06 comes from. It is helpful here to think in terms of hundredths. Before the 6% price increase we have $\frac{100}{100}$ of the given amount. Adding 6% will increase this to $\frac{100}{100} + \frac{6}{100} = \frac{106}{100}$, which equals 1.06.

So, the new price = $1.06 \times 82.1 = 87.026$p.

This rounds to 87.0p, which confirms the previous answer from the two-staged method.

EXAMPLE 4   15% reduction

This time the socks sale is rather less inviting. As you can see, the reduction now is only 15%!

> # SOCKS
> Improve your *socks* life!
> ~~£2.50~~ a pair
> ## Now 15% off!!

What is the sale price of a pair of these socks?

Solution
Again, since 15% is not easily converted into a convenient fraction, it makes sense to do this calculation on a calculator. As for Example 3, I will first do it the long-winded, two-staged way and then more directly using the one-staged method.

Stage1:   Find the price decrease
            The price decrease = $0.15 \times 2.50 = £0.375$
            This can be rounded up to the next penny, i.e. £0.38.

Stage 2:   Subtracting the price reduction
            The new price = £2.50 − £0.38 = £2.12

As was the case with Example 3, the whole process can be reduced to a single stage, by multiplying the old price by 0.85. Again, it is helpful to think in terms of hundredths. Before the 15% price decrease we have $\frac{100}{100}$ of the given amount. Subtracting 15% will decrease this to $\frac{100}{100} - \frac{15}{100} = \frac{85}{100}$ which equals 0.85.

So the new price = $0.85 \times £2.50 = £2.125$.

After rounding, this confirms the previous answer from the two-staged method.

You will need some practice at calculating percentage increases and decreases, so have a go at Exercise 6.3 now.

EXERCISE 6.3   Calculating percentage increases and decreases

(i)   A table normally sells at £42. How much will it cost with a 30% reduction?

(ii)   Check my garage bill.

Anytown Autos	
Full service repairs and parts	£186.40
VAT @17.5%	£34.62
Total (inc. VAT)	£229.02

(iii) If you earn £230 per week, which would you prefer? A rise of: **a** 6%   or   **b** £12 per week?

(iv) Your taxable earnings are £884 this month. How much of this will you have left after paying 33% in stoppages?

Persistent problems with percentages

It will be no surprise to you to be told that a lot of children's time in school takes place with one eye shut and the other staring out of the window. Many children come away from a lesson in percentages (or whatever) with only a few pieces of the jigsaw and have to somehow fill in the rest of the picture themselves. Unfortunately they don't always get it right ...

Can you spot where this child has gone wrong?

The problem is that she has started from a true fact that $10\% = \frac{1}{10}$ and built up a rule which doesn't work for any other fraction. This is probably the most common misapprehension about percentages. If you still have problems with this, the chances are that they can be traced back to a fuzziness about fractions. You may know that 20% is more than 5%. However, it is *not* the case that $\frac{1}{20}$ is more than $\frac{1}{5}$. If you think back to Chapter 04 and the idea of a fraction being a slice of cake, then imagine a cake cut into twenty equal slices. Each of these slices is a twentieth of the cake and is therefore a very small slice indeed. One fifth, on the other hand is a large slice.

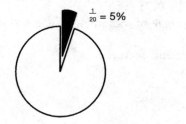

$\frac{1}{20} = 5\%$

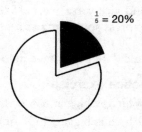

$\frac{1}{5} = 20\%$

Shopping during the sales is an opportunity to check out some of these ideas. For example, $\frac{1}{3}$ off is a better discount than 10% off. Also remember that 40% off the price of something fairly cheap like a packet of envelopes represents only a small saving in actual money, whereas the same percentage reduction from, say, the price of a house represents a huge saving.

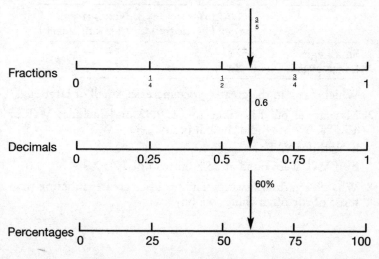

Perhaps the most important thing you need to grasp is that fractions, decimals and percentages are really the same thing. I have found that drawing three number lines one above the other is a helpful way of emphasising these connections, as shown on the previous page.

The arrows show that $\frac{3}{5}$, 0.6 and 60% have the same value.

Percentage checklist
You should now be able to:

- realize that converting to percentages makes it easier to compare fractions
- link percentages to common and decimal fractions
- convert from percentages to (simple) fractions and decimals and vice versa, e.g. $75\% = \frac{3}{4} = 0.75$
- express something as a percentage of something else, e.g. 6 is 25% of 24
- calculate percentage increases and decreases.

Practice exercise

1 Which is bigger, 8% or $\frac{1}{8}$?

2 Which is bigger, 15% or $\frac{1}{15}$?

3 If the rate of inflation drops from 5% to 4%, are prices:
   a going up? **b** coming down? **c** neither?

4 What is 20% of £80?

5 What is 10% of 20% of £80?

6 Here are some egg prices before and after a price rise.

	Old price (per half dozen)	New price (per half dozen)
Small eggs	71	75
Large eggs	84	88

Which has had the greater price increase: small or large eggs?

7 My garage bill has come to £120.93 and includes VAT at 17.5%. What would the bill be:
   a without VAT?
   b if VAT were rated at 25% instead of 17.5%?

8 Why do children's sweets tend to suffer greater inflation than most of the other things we buy?

9 Study the newspaper cuttings on page 74 and then answer the following questions.

   a From the first cutting, the final sentence ends by stating that 'typical housing costs ... would pass the £4000 mark in October'. Use the information from the rest of the article to check that this figure is correct.

   b Turning to the second article on the same page, estimate the number of residents in Campodimele who are aged between 75 and 99 years old.

Answers to exercises for Chapter 06

6.1 Fraction	Decimal fraction	Percentage
$\frac{1}{2}$	0.5	50%
$\frac{3}{4}$	0.75	75%
$\frac{7}{10}$	0.7	70%
$\frac{1}{5}$	0.2	20%
$\frac{1}{20}$	0.05	5%
$\frac{3}{5}$	0.6	60%
$\frac{3}{8}$	0.375	$37\frac{1}{2}$%

6.2 Comments in the text.

6.3 (i)   The reduction in price is 30% or $\frac{3}{10}$. Three tenths of £42 can be found on your calculator by pressing
either
   3 $\div$ 10 $\times$ 42 $=$
or
   0.3 $\times$ 42 $=$
both of which give the correct answer £12.60.
So the reduced price is £42 – £12.60 or £29.40.
(*Note*: A quicker way of doing this is to say that a 30% reduction will bring the price down to 70% of the old price. On the calculator you would press
0.7 $\times$ 42 $=$, which gives the answer £29.40.)

   (ii)   The VAT is incorrect. Using a calculator:

Press 186.40 $\times$ 0.175 $=$ to give the answer £32.62.
So I've been overcharged by £10.

Note: the direct method for checking the final bill is to press 186.40 $\times$ 1.175 $=$

(iii) 6% of £230 = £13.80, which is a bigger rise than £12.

(iv) I will have 67% of £884 left, which is £592.28.

Answers to practice exercise

1 $\frac{1}{8}$ = 12$\frac{1}{2}$% and so is bigger than 8%.
Note: You can use your calculator to convert to a percentage by pressing

1 ⌷÷⌷ 8 ⌷×⌷ 100 ⌷=⌷

2 $\frac{1}{15}$ is approximately equal to 6.7%, so 15% is bigger than $\frac{1}{15}$.

3 The annual rate of inflation measures how much average prices have risen over a year. If that rate is a positive number (such as 4% or 5%, for example), this means that prices have risen. So, even though the rate of inflation has fallen, the current rate of 4% shows that prices are still rising, but not quite as quickly as they were over the previous year.

4 20% is $\frac{1}{5}$. One fifth of £80 = $\frac{£80}{5}$ = £16.

5 10% of 20% is one tenth of 20% = 2%

   2% of £80 = £80 × $\frac{2}{100}$ = £1.60

6 The solution is summarized in the table below.

	Old price (p)	New price (p)	Price increase (p)	Percentage increase (rounded to 1 decimal place)
Small eggs	71	75	4p	$\frac{4}{71}$ × 100 = 5.6%
Large eggs	84	88	4p	$\frac{4}{84}$ × 100 = 4.8%

So, although both eggs have seen the same actual price rise (4p in each case), the small eggs have shown the greater percentage rise.

7 **a** This calculation is slightly harder than the others, as it involves working backwards after the percentage increase has been added. The story line of the solution is as follows: Let the bill without VAT be thought of as 100% and the bill with VAT (costing £120.93) as 117.5%. So we must divide the total bill by 117.5 and then multiply by 100.

i.e. £120.93 × $\frac{100}{117.5}$ = 102.92 (rounded to the nearest penny).
So the net bill (i.e. not including the VAT) = £102.92.

b To calculate the bill inclusive of VAT at 25%, we multiply
the net bill by 1.25.
£102.92 × 1.25 = £128.65.

8 When percentage increases are applied to the price of goods,
it is generally the case that fractions of pence get rounded up.
For example, if a bar of chocolate costing 36p is subjected to
a 10% increase, the true price should be 39.6p. However, as
shopkeepers cannot charge 0.6 of a penny, this price is likely
to be rounded up to 40p. This represents a loss of 0.4p to the
customer and a loss of 0.4p represents a much greater
proportion of something costing 40p than of an item costing,
say, £40. Since the things that children buy (comics, sweets,
etc.) tend to be cheap, children lose out from these rounding
losses more than adults.

9 **a** Earlier in the article, the claim was made that housing
costs would soon be 25 per cent higher than they were a
year ago. To check an increase of 25% from a starting
value of £3200, press the following on your calculator:
1.25 ⌧ 3200 ⌹
This confirms the result of £4000 mentioned in the last
sentence.
Alternatively, you might be able to do the calculation in
your head, as follows.
25% is one quarter, and one quarter of £3200 is £800.
Adding we get £3200 + £800 = £4000.

Finally, however, you also need to check that the 25%
increase related to the full one year period for which it
applied. Since the comments were made in February 1995,
the period from (the previous) April 1994 to (the next)
October 1995 covers more than one full year (actually
about 18 months), so the claim does seem justified.

b This article talks about 'more than 10 per cent' of 900
people.
Since 10% = $\frac{1}{10}$, one tenth of 900 = 90. So there are more
than 90 residents aged between 75 and 99 years old.

07 measuring

In this chapter you will learn:
- about measuring dimensions and units
- how to round numbers
- how to convert units of measure.

It is said of frogs that they sort all other animals they meet into just three catgories.

  If it is small, they eat it.
  If it is large, they run away from it.
  And if it is about their own size, they mate with it.

I think it's fair to say that, in general, humans are slightly more discriminating! Any activity which involves making judgements about the size of things can be called measuring. Although frogs may not be engaged in highly sophisticated measuring here, they *are* trying to understand the bigness or smallness of things around them.

What do we measure?

Most people tend to think of measuring as using weighing scales or a tape measure. But what sort of thing do these devices tell us about? Weighing scales tell us about weight and a tape measure about length. These types of measurement are called *dimensions*.

Dimensions of measure are measured in certain measuring *units*. For example, weight may be measured in kilograms, grams, ounces, pounds, and so on, while length may be measured in centimetres, metres, inches, miles, and so on.

There are, of course, many dimensions other than length and weight which we need to measure, for example:

  temperature, time, area, capacity, angle, volume, speed …

Exercise 7.1 will give you a chance to think about these dimensions and also about the units in which they are usually measured.

EXERCISE 7.1   Dimensions and units

Complete the table (the first two have been done for you).

Question	Dimension of measure	Likely units of measure
How heavy is your laundry?	weight	kg or lb
How long is the curtain rail?	length	cm or in
How hot is the oven?		
How far is it to London?		
How fast can you run?		
How long does it take to cook?		
How much does the jug hold?		
How big is your kitchen?		
How big is the field?		

As you see, I have included two lots of units in the examples above because both are in common usage. They are known as the metric system of units and the imperial system of units. Because many people are confused by these various units of measure, they are explained in some detail later in the chapter.

Why do we measure?

The reason we measure is that, quite simply, we live in a more complex world than a frog. Although words like 'large' and 'small' are sometimes good enough for some particular purposes ('Give me some of the *large* apples', 'I'd like a *small* helping', and so on), often we need to be more precise. Here is an example where the word 'large' proved inadequate during a national rail strike.

A very large proportion of the staff didn't show up for work.
(Union spokesperson)

A large proportion of the staff showed up for work.
(Management spokesperson)

The use of the word 'large' in these quotations is highly dubious. How large would the proportion have to be for you to consider it 'large' – 10%, 25% or perhaps 70%? ... It is interesting that both sides in the dispute have been deliberately vague about the exact figures and prefer instead to give a general impression.

Sometimes, however, a general impression is simply not good enough, and something more precise is needed. For example, you may have seen signs on the motorway advising drivers of 'large' vehicles to stop at the next emergency phone and contact the police. If you are driving a lorry, how would you know whether this referred to you? Rest assured that the small print below the sign goes on to explain that:

'Large means 11′ 00″ (3.3 m) wide or over'

The reason that we tend to measure with numbers is to help us make decisions and comparisons fairly and accurately. Careful measuring helps us bake 'the perfect cake' every time, lay well-fitting carpets with a minimum of waste, check that the children's shoes don't pinch and so on. In Exercise 7.2 you are asked to think about the measuring dimensions involved in these sorts of everyday tasks.

EXERCISE 7.2   Being aware of the dimensions of measure

Here is a list of eight common dimensions of measure.

Length (L), Area (A), Volume (V), Weight (W),
Time (T), Temperature (T°), Capacity (C) and Speed (S).

Make a note of which of them are likely to be important in the following everyday activities. (Note: there may be several dimensions involved in each activity.)

Everyday activities

• Baking a cake
• Buying and laying a carpet
• Checking the children's shoes
• Setting out on a journey in good time

How do we measure?

Measuring is basically a way of describing things. Descriptions can come in two basic forms. First, there are descriptions of *quality* and these tend to be made with words. For example, you may describe your various friends as happy, carefree, moody, thoughtful, sensitive, and so on. These are not the sorts of descriptions that easily lend themselves to being reduced to numbers. Descriptions of *quantity*, on the other hand, do involve numbers. For example, Ann is 1.59 m tall, Donal is 73 years old, Chris has 5 children, and so on. When people talk about measurement, they are usually thinking about measurement based on numbers, but not always.

Some measuring seems to fall between quality and quantity. For example, you might describe something as being large or small, fast or slow, chilly or warm. These are ways of indicating whereabouts on some sort of scale (respectively they refer to size, speed and temperature). Yet, although they make no mention of numbers, these descriptions are a sort of measure. Such words can be *ranked* into a meaningful order and they then produce what is called an *ordering* scale. On the other hand, words which describe, say, an emotion or a colour, do not normally relate to a useful scale. Thus, you can't say that 'curious' is bigger than 'excited' or that 'red' is more than 'blue'. Such words are simply descriptions. Exercise 7.3 will give you practice at using an ordering scale.

EXERCISE 7.3   Using and ordering scale

Here are five words used in describing how *likely* something is to happen:

likely, impossible, doubtful, certain, highly improbable

These descriptive words can be written in order of likelihood, from least likely to most likely, thus:

impossible, highly improbable, doubtful, likely, certain

 least likely

 most likely

Now, here are some for you to do.

Rank the following sets of words into useful ordering scales.

1 These five words are used in describing ways of travelling on foot:

jog, stop, sprint, walk, amble

2 These are the developmental stages that babies usually go through:

walk, lie, sit up, stand, roll over

3 These words are often written in sequence on an electric iron:

wool, linen, silk, cotton, rayon

To summarize, then, measuring can take the following three basic forms:
• words alone
• words which can be ranked in order
• numbers.

The types of measuring scale which these three approaches use are:
• words
• ordering scale
• number scale.

Although all three types of scale are helpful in providing an interesting variety of descriptions and comparisons, it is the third of these, measuring with numbers, which is the most important in mathematics.

How accurately should we measure?

The accuracy with which we measure depends entirely on what and why we are measuring. A brain surgeon and a tree surgeon have different needs for accuracy when sawing up their respective 'patients'. A nurse weighing out drugs will exercise greater care and precision than a greengrocer weighing out potatoes. A calculator can sometimes give a false sense of the accuracy of an answer. As the example below shows, it may give a result showing eight-figure accuracy but the numbers on which the calculation was performed may be only approximate. Suppose you wish to replace the fence in your garden. The length of fencing needed is, say, 21 m and each panel of fencing is 1.8 m in length.

Pressing 21 ⌷÷⌷ 1.8 ⌷=⌷ on your calculator will probably produce the answer 11.666666 (for reasons that will be explained shortly, on some calculators the answer will be shown as 11.666667). For this sort of calculation, it is plainly silly to give an answer to eight figures. If the last six digits of your answer are either dubious or unnecessary, then dispose of them. However, you have to be a bit careful how you do this. Numbers can be shortened so that you finish with a suitable number of digits (say three). This is called giving your answer 'correct to three significant figures' (or 'to 3 sig. figs.', for short). With the above example, calculating the number of panels of fencing requires that you buy a whole number of panels, so the answer will be given correct to two significant figures. In this case 11.666666 would be rounded up to 12 panels. *Note*: You would still need 12 panels even if the answer on the calculator was 11.333333!

This process of simplifying unnecessarily accurate measurements to a near approximation is called *rounding*. Some example are given in Table 7.1 below.

Measurement	Rounded to 3 sig. figs.
4.18345926	4.18
371.41429	371
0.0142419	0.0142
74312.692	74300
11.6666	11.7

table 7.1 rounding to 3 significant figures

Notice that the third and fourth examples in Table 7.1 have produced answers which contain not three but five figures. However, the two zeros at the beginning of 0.0142 and the two zeros at the end of 74300 are not considered to be significant figures. They are only there to give the overall magnitude of the number. It makes sense to do this as otherwise the number 74312.692 would be rounded to 743, which is clearly nonsense!

The last example in the above table, 11.6666, is different from the others in the following respect. As you can see, its third digit has been *rounded up* from a 6 to a 7. The clue to why it has been rounded up can be found by looking at the fourth digit in the original number: the 6. Since it is bigger than 5, the 6 in the tenths column is *rounded up* to a 7. And this is the reason that some calculators produce the answer 11.666667 for the fence panel calculation. Such calculators have been designed so that they round up the final digit displayed when the next digit would have been a 5 or greater. In order to be able to do this, these calculators need to process their calculations to greater accuracy than the eight figures that they display, which accounts for why they tend to be slightly more expensive than the calculators which don't round. Exercise 7.4 gives you practice at rounding.

By the way, don't worry if this explanation of rounding sounds confusing – it is easier to do than to read about!

EXERCISE 7.4   Rounding practice

Round the following numbers to four significant figures.

	Number	Answer to 4 sig. figs.
1	4124.7841	4125
2	38.4163	
3	291.7412	
4	39042.611	
5	39048.619	
6	38.4131	
7	446.982	
8	0.142937	
9	1317.699	
10	3050.1491	

There are many practical situations where careful measurement is essential – for example, dress-making, carpentry, weighing out parcels to calculate the cost of postage, and so on. However, in other situations, an *estimate* based on experience and common sense is often good enough. For example, when returfing a lawn, you may wish to measure its area fairly accurately using a tape measure, but if you decide to seed it, simply pacing it out to estimate the area may be sufficient. Estimation is a skill which greatly improves with practice. I sometimes find it helpful to imagine everyday objects of a standard size to help me make an estimate. For example:

Estimate	*Helpful image*
• estimating height or distance	– a door is roughly 2 metres high
	– a running track is 400 metres around
• estimating capacity/volume	– a standard milk bottle holds one pint
• estimating weight	– a bag of sugar weighs 1 kg
• estimating air temperature	– typical winter temperatures 0°C – 10°C
	– summer temperatures 20°C – 30°C
	– spring/autumn temperatures 10°C – 20°C

And now, as promised earlier in the chapter, we turn to the units that are used in measuring.

Imperial and metric units

Until about 1970, measurement in the UK was largely done with imperial units. Since then, however, the British population has at last owned up to the fact that they have the same number of fingers and thumbs as the rest of the world, and have 'gone decimal'. The *decimalization* of money in 1971 was carried out quickly and effectively. As a result, most people mastered the new coinage within days. It was also intended that the familiar imperial units of length, weight and capacity be phased out within a few years. This change, called *metrication*, was to have swept away the most familiar of the measuring units – feet, inches, yards, pounds, stones, pints, gallons and so on – in favour of metres, kilograms, litres and the like. Indeed, during the 1970s, many children learnt only the metric units in school on the assumption that the old imperial units would soon be six feet (sorry, 1.83 metres) under. Unfortunately, however, the change-over was so half-hearted that, at the present time of writing, we are still regularly using both systems (and having fun trying to convert from one to the other!). Since the 1980s, children have been taught both systems in school.

Conversions between metric and imperial units tend to involve rather awkward numbers. For example, there are 'about' 39.370078 inches in one metre! Not surprisingly, a number of half-baked approximations have appeared like the metric yard, the metric foot and even the metric brick. Have a look at the table below and you will see just how 'approximate' some of the approximations are.

Unit	'True' value
Metric yard = 39 in	39.370078 in
Metric inch = 2 cm	2.54 cm
Metric foot = 30 cm	30.48 cm
Metric mile = 1500 m	1609.344 m

Table 7.4, given near the end of this chapter, summarizes most of the metric and imperial units that you are likely to need. I will explain how to use it by focusing on the most basic measure of all – length.

Metric units
millimetre – 10 → centimetre – 100 → metre – 1000 → kilometre
  (mm)   (cm)   (m)   (km)

Imperial units
inch – 12 → foot – 3 → yard – 1760 → mile
 (in)   (ft)   (yd)

table 7.2 measuring length in imperial and metric units

The numbers on the arrows tell you how to convert from one unit to another. Thus, there are 10 mm in 1 cm, 100 cm in 1 m, and so on. If you want to know how many mm are in 1m, then *multiply* the two numbers 10 and 100 (i.e. there are 1000 mm in 1 m). It will help you understand and remember the metric units when you realize that:
• for each dimension there is a *basic unit* – the basic unit for length is the metre
• all the other units get their name from the basic unit: e.g., because *centi-* means one hundredth ($\frac{1}{100}$), then a *centimetre* is one hundredth of a metre.

Table 7.3 will help you work out the others.

MILLI- one thousandth ($\frac{1}{1000}$)
CENTI- one hundredth ($\frac{1}{100}$)
DECI- one tenth ($\frac{1}{10}$)
KILO- one thousand (1000)

table 7.3 metric prefixes

Converting *between* metric and imperial units is a little trickier. If you don't need to be too accurate, it is helpful to remember that a twelve-inch ruler is almost exactly 30 cm long. Dividing 30 by 12, it follows that one inch is roughly equal to 2.5 cm. When you need to be more accurate, use the conversion 1 inch = 2.54 cm, and also use a calculator!

Area

Length is a *one-dimensional* (1-D) measurement because it involves only one direction. Problems involving surfaces (size of paper, carpets, curtain material, lawns, ...) are *two-dimensional* (2-D). Sometimes we describe area simply by stating the length and the breadth. For example, curtain material is bought by the metre (length) but we also need to know that the roll is 1 m 20 cm wide. If you are buying paint, on the other hand, the instructions on the tin may say something like: '... contents sufficient to cover 35 m²'. The unit described as a 'm²', or a 'square metre' is the basic metric unit of area. It means exactly what it says. One m² is the area of a square, 1 m by 1 m, as illustrated below.

1 m

The area of this 1 m × 1 m square is equal to 1 m².

There should be enough paint in the tin to cover 35 of these squares.

1 m

Similarly, a ft² (square foot) is the area of a square 1 ft by 1 ft, and so on for the various other units of area. However, it's all not quite as easy as it sounds. Have a go at Exercise 7.5 now and see if you can avoid the traps that people often fall into.

EXERCISE 7.5   Area traps

a How many square feet (ft²) are there in one square yard (yd²)?
b How many cm² are there in one m²?
c What is the area of this rectangle?
d If you double the dimensions of this rectangle (i.e. double the length *and* the breadth), what do you do to the area?

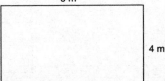

6 m

4 m

Volume

If you ask most people about the word 'volume', they will tell you that it is the knob on the TV set which makes it go loud and quiet. 'Volume', as used in mathematics, is rather different. It describes an amount of space in *three dimensions* (3-D). If you think of a box, its volume will depend on the three dimensions length, breadth and height.

Unlike the words 'length' and 'area', 'volume' is not a word in very common everyday usage. We tend, instead, to use terms like:
How big is the brick? or
What is the *size* of the box?

However, the trouble with words like 'big' and 'size' is that they don't necessarily refer to volume. In fact they can be called upon to describe any of a number of dimensions. For example, the size of a pencil might mean its length. The size of a piece of paper might mean its area. The size of a bag of sugar might even mean its weight, and so on.

People who work in the building trade become skilled at estimating amounts of earth and concrete. They usually measure these volumes in so many 'cubes'. A 'cube' usually refers to cubic metre or a cubic foot. A cubic metre is the amount of space taken up by a cube measuring 1m by 1m by 1m.

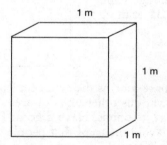

Similarly a cubic centimetre (cc) is the amount of space taken up by a 1 cm cube.

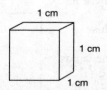

Another measure which deals with three dimensions is capacity. The difference between capacity and volume is that capacity describes a container and is a measure to show how much the vessel holds. For example, we talk about the capacity of a saucepan, bucket, bottle, etc. The units of capacity, which are included in Table 7.4, are normally only used with liquids.

Have a look now at Table 7.4, which shows some of the most common metric and imperial units for length, area, volume, capacity and weight.

Metric units	
Length	millimetre (mm) – 10 → centimetre (cm) – 100 → metre (m) –1000 → kilometre (km)
Area	centimetre² (cm²) – 10 000 → metre² (m²) – 10 000 → hectare
Volume	centimetre³ (cm³) – 1 000 000 → metre³ (m³)
Capacity	millilitre (ml) – 1000 → litre (l)
Weight	gram (g) – 1000 → kilogram (kg) –1000 → tonne

Imperial units	
Length	inch (in) – 12 → foot (ft) – 3 → yard (yd) – 1760 → mile
Area	inch² (in²) – 144 → foot² (ft²) – 9 → yard² (yd²) – 4840 → acre – 640 → square mile
Volume	in³ – 1728 → ft³ – 27 → yd³
Capacity	fluid ounce – 20 → pint – 2 → quart – 4 → gallon
Weight	ounce (oz) –16 → pound (lb) – 14 → stone – 112 → hundredweight (cwt) – 20 → ton

table 7.4 metric and imperial units

You will probably need practice at using these tables, so do Exercise 7.6 now.

EXERCISE 7.6   Getting familiar with the units of measure

a How many millimetres are there in a metre?
b How many centimetres are there in a kilometre?
c How many grams are there in a tonne?
d How many inches are there in a mile?
e How many square inches are there in a square yard?
f How many ounces are there in a ton?

Unfortunately it isn't enough to be able to convert units within the metric or the imperial system separately. Sometimes it is necessary to convert between the two. Table 7.5 shows some of the most common conversions between metric and imperial units.

Accurate conversion		Rough 'n' ready conversion
Length	1 m = 39.37 in	1 metre is just over a yard (a very long stride)
	1 in = 2.54 cm	1 inch = $2\frac{1}{2}$ cm
Capacity	1 litre = 1.76 pints	1 litre = $1\frac{3}{4}$ pints (a large bottle of orange squash)
	1 gallon = 4.54 litres	1 gallon = $4\frac{1}{2}$ litres
Weight	1 kg = 2.2 pounds	1 kg = just over 2 lb – a bag of sugar
	1 pound = 0.454 kg = 454 g	1 lb = just under $\frac{1}{2}$ kilogram
Speed	100 kmph = 62.1 mph	8k mph = 5 mph
	100 mph = 161 kmph	

table 7.5 converting between metric and imperial units

Again, it would be a good idea to practise some of these conversions now, so have a go at Exercise 7.7.

EXERCISE 7.7 Practice exercise

Use Tables 7.4 and 7.5 (and a calculator where appropriate) to answer the following:

a Which of these would be a reasonable weight for an adult? 60 kg, 600 kg, 6 kg

b Is it better value to buy a 25 kg bag of potatoes or a 56 lb bag for the same money?

c What is the height of your kitchen ceiling from the floor, in metres?

d Some French roads have 90 kmph speed limits. What is this roughly in mph?

e Is $\frac{1}{2}$ litre of beer more or less than a pint?

f If we bought milk by the $\frac{1}{2}$ litre, how many bottles would you have to buy to have roughly 7 pints?

Finally, here is a checklist of the basic skills of measuring which you will need.

Measuring checklist

• be familiar with the common measures
  (length, weight, area, volume, capacity, time, speed, temperature)

- use and understand standard units of these measures (centimetre, kilogram ...)
- estimate lengths, weights and so on, in terms of these units, to know when a measurement is about right and to be aware what sort of accuracy is appropriate
- use various measuring instruments (tape measure, ruler, weighing scales, balance, measuring jug, thermometer, clock)
- be aware of composite units (miles per hour, price per gram, and so on).

As a footnote, I feel that I should raise the issue of my use of the word 'weight' throughout this chapter. Strictly speaking, I should talk about 'mass' rather than weight. The weight of an object is a measure of the force of gravity acting on it and this will vary depending on where the object is in relation to the earth. Mass is the 'amount of matter' which the object contains and, wherever its position, this will not vary. Most scientists feel that this distinction is critical but, provided your mathematics is conducted mostly on the Earth's surface, I wouldn't let it worry you too much!

Summary

This chapter started by looking at the 'what', 'why' and 'how' questions of measurement.

What do we measure?	length, area, volume, capacity, weight, time, temperature, angle, speed, ...
How do we measure?	– using words alone – using words ranked in order (an ordering scale) – using numbers (a number scale)
Why do we measure?	To help make decisions and comparisons.

The final part of the chapter looked at the common metric and imperial units of measure and at how we can convert within and between the two systems.

Answers to exercises for Chapter 07

7.1 *Question*

Question	Dimension of measure	Likely units of measure
How heavy is your laundry?	weight	kg or lb
How long is the curtain rail?	length	cm or in
How hot is the oven?	temperature	degrees (°C or °F)
How far is it to London?	length/distance	km or miles
How fast can you run?	speed	km hr or mph
How long does it take to cook?	time	minutes
How much does the jug hold?	capacity	cc, pint or fl oz
How big is your kitchen?	volume	m^3 or ft^3
How big is the field?	area	hectare or acre

7.2 *Everyday activities*

Measuring dimensions

- Baking a cake — W; T; T°; C
- Buying and laying a carpet — L; A
- Checking the children's shoes — L; C
- Setting out on a journey in good time — L; T; S

7.3 stop, amble, walk, jog, sprint
lie, roll over, sit up, stand, walk
rayon, silk, wool, cotton, linen

7.4

	Number	Answer to 4 sig. figs
1	4124.7841	4125
2	38.4163	38.42
3	291.7412	291.7
4	39042.611	39040
5	39048.619	39050
6	38.4131	38.41
7	446.982	447.0
8	0.142937	0.1429
9	1317.699	1318
10	3050.1491	3050

7.5 a $9 \ ft^2 = 1 \ yd^2$

b $10\ 000 \ cm^2 = 1 \ m^2$

c Area $= 6 \ m \times 4 \ m = 24 \ m^2$

d doubling the length and the breadth makes the area four times as big.

7.6 a Number of millimetres in a metre = $10 \times 100 = 1000$

   b Number of centimetres in a kilometre = 100×1000 = $100\,000$

   c Number of grams in a tonne = $1000 \times 1000 = 1\,000\,000$

   d Number of inches in a mile = $12 \times 3 \times 1760 = 63\,360$

   e Number of square inches in a square yard = $144 \times 9 = 1296$

   f Number of ounces in a ton = $16 \times 14 \times 112 \times 20 = 501\,760$.

7.7 a 60 kg would be a reasonable weight for an adult.

   b A 25 kg bag weighs $25 \times 2.2 = 55$ lb. So, buying a 56 lb bag for the same money is a slightly better deal.

   c A typical height of a kitchen ceiling from the floor is roughly $2\frac{1}{2}$–3 m.

   d $90 \times 5 \div 8 = 56$ mph, roughly.

   e 1 litre = 1.76 pints. So, since $\frac{1}{2}$ litre = $\frac{1.76}{2}$ = 0.88 pints, this is less than one pint.

   f 7 pints = $\frac{7}{1.76}$ = approximately 4 litres, or 8 half litre bottles.

08

statistical graphs

Open a newspaper or watch the news on TV and you will be expected to make sense of a range of charts and graphs and to process statistical facts and figures. For example, here is a statistical fact.

> Did you know that eight out of ten advertisers are prepared to mislead the public a little in order to sell their product? Furthermore, the other two are prepared to mislead the public a lot!

One of the troubles with statistics is that there is such scope for deception. For example, I just made up the figures quoted above out of my head. But writing them 'in black and white' somehow seems to lend credibility to so-called 'facts'.

Deception can occur not only through the quoting of incorrect information. Equally common is 'dirty dealing' by means of the incorrect display of correct information. This chapter deals with the charts and displays that are most commonly used and misused in the media – barcharts, piecharts, line graphs and tables – as well as scattergraphs. It provides examples of where they are used, what they mean, how they are interpreted and how they are sometimes misused to create a false impression.

Barcharts and piecharts

Barcharts and piecharts are useful when we want to compare different categories. Barcharts (sometimes called block graphs) consist of a set of bars set either vertically or horizontally. The height (or length) of each bar is an indication of its size. Piecharts also allow different categories to be compared but here the size of each item is represented by the size of its slice on a pie. This chapter examines different types of barcharts and piecharts.

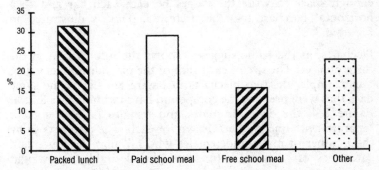

figure 8.1 vertical barchart showing the lunchtime meals of pupils

Source: *Social Trends 25*, Figure 3.13, CSO

The main strength of a barchart is that columns placed side by side or placed one above the other are easier to compare. So, in Figure 8.1, for example, you can see at a glance that the most popular form of lunchtime meal of pupils for the year in question was a packed lunch. You can also see that roughly twice as many children took a packed lunch as ate a free school meal.

Sometimes barcharts can be used to show categories from more than one source on the same graph. The most common way of doing this is to use a compound barchart, as shown in Figure 8.2.

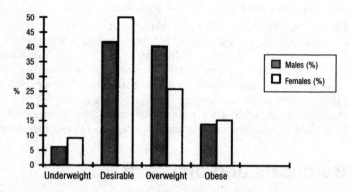

figure 8.2 compound barchart showing body mass by gender

This compound barchart shows two for the price of one – both male and female data are represented on the same graph.

Sometimes you may wish to draw a barchart where the category names are rather long. With a vertical barchart there simply isn't enough space to write the names beneath each bar and so a horizontal barchart may be preferred. This is illustrated in Figure 8.3.

Piecharts, as the name suggests, shows the information in the form of a pie. The size of each slice of the pie indicates its value. For example, the two piecharts in Figure 8.4 depict the same data that were used for the compound barchart in Figure 8.2. In Figure 8.4, the data for males and females have been kept separate, with one piechart drawn for each. The piecharts show the number of males and females who fall into the four main categories of body mass (underweight, desirable, overweight and obese).

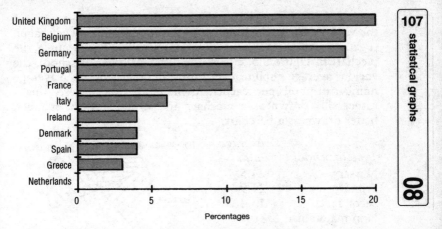

figure 8.3 horizontal barchart showing the seawater bathing areas not complying with mandatory coliform standards: EC comparison, 1993

Source: *Social Trends 25*, Figure 11.17, CSO

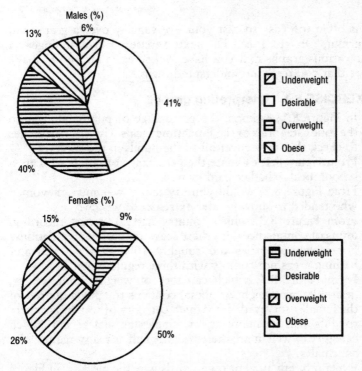

figure 8.4 piecharts showing the body mass indices of males and females

Source: *Social Trends 25*, Figure 7.6, CSO

An important feature of a piechart is that it only makes sense if the various slices which make up the complete pie, when taken together, actually represent something sensible. For example, the piechart in Figure 8.5 shows, for different types of school, the various average pupil/teacher ratios; in other words the average number of pupils per teacher. As the figure title suggests, this is rather silly drawn as a piechart and would have been much better drawn as a barchart.

Type of school	Average ratio
Nursery	21.5
Primary	21.9
Secondary	15.4
Non-maintained	10.4
Special	5.8

figure 8.5 a silly piechart showing pupil/teacher ratios by type of school, UK

Source: *Social Trends 25*, Figure 3.9, CSO

It is often too easy to cast your eye vaguely over a graph and murmur, 'Oh, yes, I see.' The next exercise asks you to linger on the various graphs that you have looked at so far and to make sure that you really do understand them.

EXERCISE 8.1   Interpreting graphs

a In Figure 8.1, estimate the percentage of pupils falling into the four categories of the lunchtime meals. Use your estimates to check that they cover all of the pupils in the survey.

b From Figure 8.1, estimate the percentage of pupils who ate a school meal, whether paid or free.

c From Figure 8.2, would you say that it was men or women who tended to show greater extremes of weight?

d From Figure 8.3, which country had the worst record in terms of compliance with these seawater bathing regulations? For which countries did roughly 10% of their seawater bathing areas not comply with these regulations?

e From Figure 8.4, which category of weight (underweight, desirable, overweight or obese) contains roughly a quarter of the females surveyed? For which category of weight are there roughly half as many again females as males? For which category of weight are there roughly half as many again males as females?

f Explain briefly in your own words why the piechart in Figure 8.5 is silly, and make a rough sketch of what it would look like redrawn as a barchart.

Scattergraphs and line graphs

The graphs you have looked at so far have been helpful if you want to make comparisons – barcharts allow you to make comparisons based on the heights of the bars, while comparisons within piecharts are based on the relative sizes of the slices of the pie.

Sometimes, however, we wish to know how two different measures are related to each other. For example:

- Does a person's blood pressure relate to the fat intake in their diet?
- Is a child's health linked to the size of the family's income?
- How has a particular plant grown over time?
- Are lung cancer and heart disease linked to smoking? and so on.

To answer these sorts of questions, which look at two different measures together, we need a two-dimensional graph. This usually takes the form of either a scattergraph or a line graph.

The figures in Table 8.1 show the contributions to, and receipts from, the EC budget in 1993. The same information is also portrayed in the scattergraph shown in Figure 8.6. By analysing the scattergraph, it is possible to explore the relationship between contributions and receipts. For example, 'Do countries which give the most tend to receive the most?', and so on.

Country	Contributions £ billion	Receipts £ billion
Germany	14.9	5.6
France	9.0	8.2
Italy	8.0	6.8
United Kingdom	5.9	3.5
Spain	4.0	6.5
Netherlands	3.1	2.1
Belgium	1.9	1.9
Denmark	0.9	1.2
Greece	0.8	4.0
Portugal	0.7	2.6
Ireland	0.5	2.3
Luxembourg	0.2	0.3

table 8.1 contributions to, and receipts from, the EC budget, 1993

Source: *Social Trends 25*, Figure 6.21, CSO, 1995

In order to draw a scattergraph of this information, you must place one of the measures on one of the axes, one on the other and mark on each axis a suitable scale. As you can see from Figure 8.6, I have chosen to place 'Contributions' to the EC on the horizontal axis and 'Receipts' on the vertical axis. Each country is plotted as a separate point. For example, the point corresponding to Germany is shown on the extreme right of the graph. Following the down arrow to the Contributions axis, you can see that this point lines up with the value £14.9 billion. Reading across to the Receipts axis from the same point, the corresponding value is £5.6 billion.

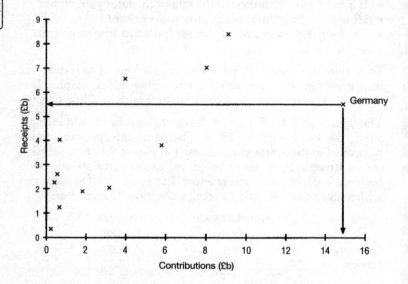

figure 8.6   scattergraph showing the relationship between net contributions and net receipts to the EU (based on the data in table 8.1)

Source: *Social Trends 25*, Figure 6.21, CSO

The pattern of points on a scattergraph helps to reveal the sort of relationship between the two measures in question. For example, in Figure 8.6 you can see that the points lie in a fairly clear pattern running from bottom left to top right on the graph. This reflects the not-too-surprising fact that countries which make small contributions (like Ireland and Luxembourg, for example) also tend to get small receipts. Similarly, the major contributors, like Germany and France, are also two of the largest receivers of money from the fund. You will probably need to spend some time consolidating your understanding of a scattergraph, so do Exercise 8.2 now.

EXERCISE 8.2   Practising scattergraphs

Do you think that countries which have a high rate of marriage also tend to have a high divorce rate? Have a look first at Table 8.2 and then at the corresponding scattergraph in Figure 8.7.

a   Check that you understand how the points have been plotted and try to match each point up to its corresponding country.

b   Why do you think there are no data for Ireland in the divorce rate column? How has the point corresponding to Ireland been shown on the graph?

c   Overall, does the scattergraph show a clear relationship between a country's marriage rate and divorce rate?

Country	Marriage rate	Divorce rate
Germany	5.6	1.7
France	4.7	1.9
Italy	5.3	0.5
United Kingdom	5.4	3.0
Spain	5.5	0.7
Netherlands	6.2	2.0
Belgium	5.8	2.2
Denmark	6.2	2.5
Greece	4.7	0.6
Portugal	7.1	1.3
Ireland	4.5	–
Luxembourg	6.4	1.8

table 8.2   marriage and divorce rates: EC comparisons, 1992 (rate per 1000 population)

Source: *Social Trends 25*, Figure 2.13, CSO

Line graphs, like scattergraphs are two-dimensional, so again we will be dealing with two measures at a time and examining the relationship between them. A line graph is one of the most common types of graph and indeed is what most people think of when we use the word 'graph'.

Figure 8.8 shows a particular type of line graph, known as a *time graph*. It is so called for the obvious reason that the measure on the horizontal axis is time.

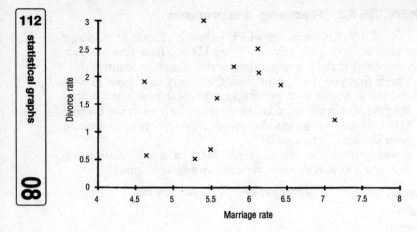

figure 8.7 scattergraph based on the data in table 8.2

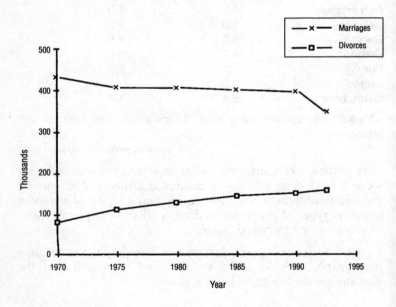

figure 8.8 a time graph showing marriages and divorces in Great Britain

Source: Adapted from *Social Trends 25*, Figure 2.14, CSO

EXERCISE 8.3 Interpreting line graphs

a How have levels of marriage and divorce altered in Great Britain over the 22 years between 1970 and 1992?

b Estimate the number of marriages in 1992 in Great Britain.

c Notice that the points marked on the two line graphs have been taken over five-year intervals. In the original line graphs (printed in the publication *Social Trends 25*), the points were based on data taken in one-year intervals. How do you think the size of the interval might affect the overall shapes of the line graphs?

Misleading graphs

If you read through newspapers and magazines, it isn't difficult to spot graphs which are misleading. The graph below contains a fine collection of disasters! See how many you can spot.

Unemployment graph for the North

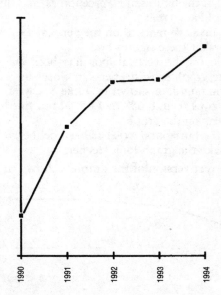

figure 8.9 spot the errors on this graph!

In order to make sense of this graph and to see some of the distortions it contains, you really need to have a look at the data from which it was drawn. These are given in Table 8.3.

Year	Rate (%)
1990	8.6
1991	10.6
1992	11.2
1993	11.2
1994	11.7

table 8.2 unemployment rates (%) for the North of England (1990 to 1994)

The graph shown in Figure 8.9 certainly looks dramatic. However, although unemployment rates in the North of England did rise over this period, the rise was not as dramatic as suggested by this representation. There are a number of errors and misleading features of the graph. Let's go through them in turn.

- The *title* is not very helpful. There is no clear explanation of what region the graph refers to ('the North' could refer to anywhere), or to what is being measured (the title should state that these are percentages).
- The *axes* are not labelled. The vertical axis should show clearly that the figures are 'Percentages' and the horizontal axis should say 'Year'.
- There is no *scale* marked on the vertical axis, so you have no idea what these figures are.
- The scale on the vertical axis has been cut in order to make the graph look steeper. On closer inspection of the corresponding data shown in Table 8.3, you can see that the axis runs from 8.6% to 11.7%, but this has not been made clear on the graph.
- Finally, the horizontal axis has been deliberately squashed up to make the graph look steeper.

A more correct version of the graph is shown in Figure 8.10.

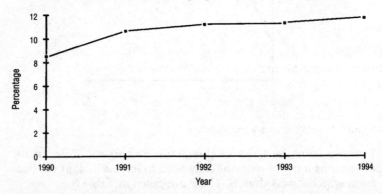

figure 8.10 line graph showing unemployment rates (%) for the North of England (1990–1994)

As you can see, now the increase is not nearly so dramatic and it is clear exactly what the figures refer to.

One final point about the scale on the vertical axis not starting at zero. In fact it is acceptable to draw a vertical scale starting from a number other than zero, provided an indication is made on the axis that this has been done. The most common method is to mark a break on the axis, as shown below.

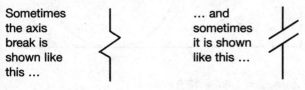

Sometimes the axis break is shown like this ...

... and sometimes it is shown like this ...

Figure 8.11 shows the same graph drawn with the vertical axis starting at 8% but with the break in the axis added to alert the reader to this potential source of confusion.

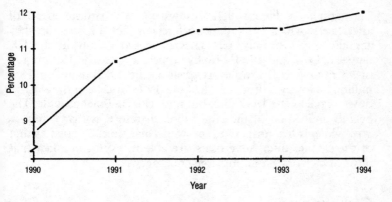

figure 8.11 a graph demonstrating an axis break

To end this section on misleading graphs, here is one of the most common types of distortion. Have a look at Figure 8.12 and see if you can spot how it might give a false impression.

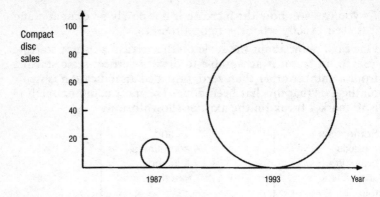

Number of discs sold in 1987 = 18.6 million
Number of discs sold in 1993 = 93 million

figure 8.12 sales of compact discs in the UK, 1987 and 1993

Source: *Social Trends 23* and *25*

The graph in Figure 8.12 shows up a favourite trick of advertisers, which is to make differences look bigger than they actually are. Certainly, CD sales increased greatly in the UK between 1987 and 1993 – by five times, in fact (the 1993 figure of 93 million is five times as great as the 1987 figure of 18.6 million). However, not only has the 1993 disc been drawn five times as tall as the 1987 disc, but it is also five times as wide. The overall impression of the larger disc, therefore, is that it has an *area* which is five times five, i.e. twenty-five times as great as that of the smaller disc. Advertisers are able to exploit the fact that most of us work on impressions, not facts!

Summary

This chapter has covered four of the most common types of graphs: barcharts, piecharts, scattergraphs and line graphs. The final section dealt with misleading graphs and a list was provided of some of the common ways in which graphs can be drawn in an unhelpful or deliberately distorted way.

Over the next few weeks, why don't you look out for some more examples of misleading graphs in newspapers and magazines. Most people who present information to the public have some sort of vested interest. You will find it helpful to ask yourself, 'What *is* the vested interest, and therefore what impression is this graph designed to convey?'

Answers to exercises for Chapter 08

8.1 a Estimates from Figure 8.1 are as follows:

Lunchtime meal	%
Packed lunch	32
Paid school meal	29
Free school meal	16
Other	23

  b The percentage of pupils who ate a school meal, whether paid or free is 29% + 16% = 45%.

  c It seems that women tend to show greater extremes of weight. This conclusion can be drawn from looking at the 'Underweight' and 'Obese' bars, which are both taller for females than for males.

  d The United Kingdom seemed to have the worst record in this respect.
  The two countries for which roughly 10% of their seawater bathing areas failed to comply were Portugal and France.

  e Roughly a quarter of the females (actually 26%) fell into the 'Overweight' category.
  Half as many again females as males (9% compared with 6%) fell into the 'Underweight' category.
  Roughly half as many again males as females (40% compared with 26%) fell into the 'Overweight' category.

  f The piechart in Figure 8.5 fails to meet a basic condition of a piechart in that the complete pie doesn't represent anything meaningful. In this case, the complete pie corresponds to the sum of the various average ratios in the different types of school, and who cares about that! The data would be more helpfully drawn as a barchart, as shown in Figure 8.13.

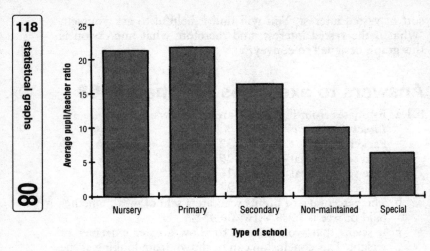

figure 8.13 vertical barchart showing average pupil/teacher ratios by type of school

8.2 a No comments.
  b Divorce was illegal in Ireland in 1992 so there can be no official divorce rate. The point corresponding to Ireland has been ignored on the scattergraph.
  c There is little evidence of any clear pattern in these points linking divorce and marriage rates.
8.3 a Marriage levels have steadily fallen while divorce levels have done the reverse over the period.
  b The number of marriages in 1992 in Great Britain is estimated to be roughly 360 thousand.
  c When graphs are plotted based on data taken at one-year intervals, as opposed to every five years, there are likely to be more subtle changes in direction. My graphs given in Figure 8.8 are actually rather crude, each being constructed by joining up six points with straight lines. However, supposing that, say, in 1987, there was a sudden blip in the graph, this would simply not have shown up on Figure 8.8.

09

using a formula

In this chapter you will learn:
- why algebra might be useful
- some of the rules of algebra
- how algebra can be used to prove things
- about spreadsheets with a computer.

Is algebra abstract and irrelevant?

For many students, the arrival of algebra in their school lessons was the point at which they felt they parted company with mathematics. Algebra has a reputation of being hard, largely because many people see it as abstract and irrelevant to their lives.

Let us first consider whether algebra is abstract. The simple answer to this charge is, 'Yes it is!'. Algebra is certainly abstract, for that is the point of algebra. The word 'abstract' means 'taken away from its familiar context'. The reason that algebra is such a powerful tool for solving problems is that it enables complex ideas to be reduced to just a few symbols. Naturally, if algebra is to be useful to you, you need to understand what the symbols mean and how they are related. Assuming this is the case, expressing something as a brief mathematical statement (which might be a formula or an equation) allows you to strip away the details, to forget about the context from which it was taken and focus on the essential underlying relationship. Of course, there are often situations where you *don't* want to strip away the context (in questions of human relationships, for example). Clearly you wouldn't wish to use algebra for such problems.

Next, let's examine the charge of algebra being irrelevant. Most people believe that they never use algebra. Yet in many jobs, particularly, say, in medicine and engineering, formulas are crucial for converting units, calculating drug dosages, setting machines correctly for different tasks, and so on.

Increasingly, large organizations and government institutions use formulas for deciding on and describing their funding arrangements. To take the example of education, a formula is used to define how much money is allocated to secondary and primary school budgets on the basis of the number and age of pupils. There are many questions that immediately arise. For example:

- Is it a fair way of allocating money?
- Is it right that the *weighting* (i.e. the relative amount) for secondary age children is much greater than for primary age children or should all children be allocated the same amount, regardless of age?
- How can you find out what the relative weightings are for children of different ages?
- Who should decide these sorts of question and can interested parents and teachers enter the debate?

The point I wish to make in introducing this chapter is that people can debate this question *only if they understand what a formula is saying*. If you don't understand basic algebra, other people will be making such decisions for you and you will have no idea whether or not they are acting in your best interests.

Before starting to examine any formulas, we begin by looking at some of the basic features of algebra – how it is used as a shorthand way of expressing something, and some of the conventions regarding how it is written.

Algebra as shorthand

There are many situations in everyday life where it is convenient to adopt a shorthand – usually in the form of abbreviations or special symbols – in order to speed things up. I am aware that, for many people, the symbols in algebra seem to cause confusion rather than be an aid to efficiency. However, the idea of using a shorthand isn't just confined to mathematics.

Some examples are given in Exercise 9.1 for you to interpret. Then, in Exercise 9.2 you are asked to examine some mathematical shorthands.

EXERCISE 9.1 Shorthands in everyday life

EXAMPLE	*Source*	*Meaning*
a det. hse, lge gdns, gd decs, FCH . . .		
b K1, P1, M1, C4F, K2, . . .		
c PAS, MOT, fsh, good runner		
d NYWJM seeks same with view to B&D, S&M, etc.		

See if you can identify the source of these shorthands and what they mean?

EXERCISE 9.2 Shorthand in mathematics

Here are some mathematical sentences. Rewrite each one in mathematical shorthand. The first one has been done for you.

	Longhand	*Shorthand*
a	Three multiplied by two and a half	$3 \times 2\frac{1}{2}$
b	The sum of twelve, and four and three quarters	
c	The sum of the squares of three and four	
d	Four times the difference of nine and three	
e	Five plus four, all divided by the product of five and four	
f	The number of inches, I, is found by multiplying the number of metres, M, by 39.37.	

As you can see from these examples, mathematics is full of shorthand notation. For example:

- instead of writing numbers out as words, 'one, two, three, and so on', we have saved time by inventing the numerals, '1, 2, 3, etc.'
- rather than say 'multiplied by' or 'added to', we use the symbols $\times$ and $+$
- squaring is represented by writing a small two above and to the right of the number or letter that is being squared. Thus, five squared, or 5 times 5, can be written as 5×5 or as 5^2.

More examples of notation will be explained in the next section. For now, let us focus on part **f** of Exercise 9.2, as it demonstrates a key feature of algebra.

The solution to this example, which is given on page 131, is repeated below for convenience.

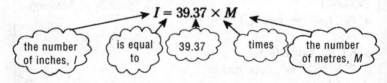

Before reading on, make sure that you can use this formula. For example, suppose you have just bought curtain material with a drop (the *drop* is typically the distance from curtain rail to the window sill) of 1.20 metres and you want to know what that is in inches. Simply replace the M in the formula by 1.20 and calculate the corresponding value of I, as follows:

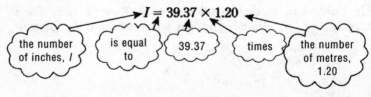

Using a calculator, the answer is 47.244 inches, a result which you might round up to 48 inches. (As an aside, rounding *up* may be appropriate here for two reasons. Firstly, if you are buying something like curtain material or a carpet, it is always better to have a little bit too much than to be a little bit short. Secondly, curtain material tends to come in various 'standard lengths' and in imperial units, 48 inches happens to be one of these standard lengths.)

Now give some thought to how the formula, $I = 39.37 \times M$, has been written.

First, notice that I have introduced the abbreviations = and × to save the trouble of writing out the words 'equals' and 'times'. You may feel that this shorthand has reduced the formula to its bare essentials, but is actually possible to dispense with the × altogether and write the formula even more briefly, as follows:

$$I = 39.37M$$

This demonstrates an important convention in algebra, namely that writing two letters together, or a number and a letter together, implies that they are multiplied. For example:

'*ab*' means *a* times *b*
'*4y*' means 4 times *y*
'*1.76L*' means 1.76 times *L*

and so on.

A second aspect of the formula worth noting is that I have used the letters *I* and *M* to represent, respectively, the number of inches and the number of metres. In algebra, the letters which we happen to choose to represent numbers are quite arbitrary. Thus, I could have written the formula as, say, $Y = 39.37X$, with *Y* representing the number of inches and *X* the number of metres. However, it is usually a good idea to relate each letter to the quantity that it represents as this will help you to remember what the various letters stand for. For that reason, I used the initial letters of Inches and Metres, i.e. *I* and *M*, in this formula.

The most common letters used in basic school algebra tend to be:

$$x, y, a, b \text{ and } n.$$

There is no obvious reason for this choice, with the possible exception of the *n*, which can be thought of as representing some unknown *n*umber.

The next section deals with formulas in practical contexts and how to use them to do calculations.

Calculating with formulae

Drug dosages need to be carefully calculated and measured out.
Giving too little of the drug means that the patient doesn't get
the full benefit, but giving too much could be highly dangerous.
The problem is greatly complicated when the drug is to be
administered to a child, because clearly a dose that would be
suitable for an adult would be too much for a young child.
There needs to be a way of adjusting the dosages depending,
perhaps, on the body weight or the age of the child concerned.
One such formula, based on the child's age, is as follows.

$$C = D \times \tfrac{A}{A+12}$$

where C represents Child dosage, D represents adult Dosage
and A is the Age of the child.

Written out in longhand, this formula means the following:

$$\text{Child dosage} = \text{Adult dosage} \times \tfrac{\text{Age}}{\text{Age}+12}$$

EXAMPLE 1

Let us take an example where the adult dose of cough medicine
is 10 mg of linctus. What would be the appropriate dosage for a
child of 6 years of age?

Solution

First let us write down what we know.

$D = 10, A = 6$

With these numbers to hand, we are ready to calculate the child's
dose, using the formula.

$C = D \times \tfrac{A}{A+12} = 10 \times \tfrac{6}{6+12} = 10 \times \tfrac{6}{18} = 3\tfrac{1}{3}$ mg

---

Now here are some to try for yourself.

EXERCISE 9.3   Calculating dosages

Using the formula $C = D \times \tfrac{A}{A+12}$, calculate the dosages for the
following situations.

a An adult prescription of a certain drug is 24 micrograms (µg).
What would be an appropriate dose for a child aged 10
years?

b An adult prescription of another drug is 200 µg. What would
be an appropriate dose for a child aged 4 years?

We now move on to another formula, this time for converting
temperatures. Temperatures in degrees Celsius can be converted
to degrees Fahrenheit with the following formula.

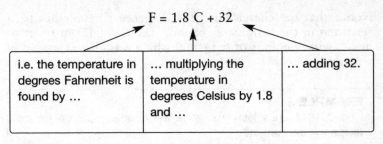

$$F = 1.8\ C + 32$$

i.e. the temperature in degrees Fahrenheit is found by …	… multiplying the temperature in degrees Celsius by 1.8 and …	… adding 32.

Here is an example of the formula in operation.

EXAMPLE 2 Oven temperatures

A typical cooking temperature for an oven is 180°C. What is this in °F?

Solution

Applying the formula:

The temperature in degrees Fahrenheit, F = 1.8 × 180 + 32

= 356°F

Now here are some for you to try.

EXERCISE 9.4   Temperature conversion

a A warm summer day's temperature would be something like 30°C. What would this be in degrees Fahrenheit?

b The boiling point of water is 100°C. What is this in °F?

c There is only one temperature which is the same in degrees F as in degrees C. Try to find it.
(*Hint*: It is a temperature well below freezing point.)

The next example of a formula is concerned with calculating a 'phone bill. Telephone bills are usually calculated each quarter on the basis of a fixed sum for the rental of the line plus a variable cost based on the calls you make. For example, my last bill of £97.61 was made up of a rental of £20.16 and a further 4.20 pence per unit used. The formula for this can be written as follows.

$$C = 20.16 + 0.042U$$

where C is the Charge in pounds and U is the number of Units used.

Notice that the charge rate of 4.20 pence per unit has been rewritten in the formula in pounds (i.e. as 0.042) in order to match with the units of the rental which was also expressed in pounds.

EXAMPLE 3
I used 1844 units last quarter. Is the quarterly charge for my telephone bill correct?

Solution
The charge is calculated as follows:
$C = 20.16 + (0.042 \times 1844) = £97.61$ (rounded to the nearest penny).
This confirms the bill which I received as being correct.

Exercise 9.5 gives you the opportunity to try some of these for yourself.

EXERCISE 9.5   More bills
Calculate the quarterly charge for a household which used:
a 944 units
b 3122 units.

Proving with algebra

This final section looks at an aspect of mathematics close to the hearts of mathematicians – the idea of proof. Without algebra, proving that a mathematical result is true is quite difficult. It is often easy enough to show that the result is true for several *particular* numbers but it is quite a different matter to say that you *know* it is true for *all possible numbers*. For example, is it the case that adding two odd numbers always produces an even answer?

We could take some examples and see if it works. Thus:

$3 + 7 = 10$, which is even
$5 + 11 = 16$, which is even
$23 + 15 = 38$, which is even
$111 + 333 = 444$, which is even.

So, it does seem to be true, but have we proved it? Certainly not! Checking only four examples does not constitute a proof. But what if I produced another twenty examples, or a hundred, or even a million? That would be quite impressive, but unfortunately providing lots and lots of special cases would not cut much mustard with a mathematician. Why, then, is it so hard to prove a numerical result to be always true? The reason is that you can't try all the infinite number of possible cases, and it would only take one of the ones you didn't get round to checking to be wrong to blow your theory apart. While arithmetic is useful for doing calculations with *particular* numbers, algebra is needed for making *generalizations*. The mathematical proof of this generalization (that the sum of two odd numbers is always even) is outlined and explained below. But before launching in to it, you need to spend a few minutes thinking about how we might represent even and odd numbers algebraically.

An aside on even and odd numbers

If we think of whole numbers as represented by, say, the letter K, then we can write even numbers as $2K$. You can check this out by giving K any value you wish to think of. For example:

when $K = 3$, $2K = 6$, which is even.
when $K = 8$, $2K = 16$, which is even.
when $K = 13$, $2K = 26$, which is even.
when $K = 50$, $2K = 100$, which is even.

and so on.

The reason we know that $2K$ is always even is that it contains a factor 2, which is essentially what an even number is.

Similarly, if we represent the whole numbers by, say, the letter L, any odd number can be written with the formula $2L + 1$. Again, let's take a few examples to check this out.

when $L = 3$, $2L + 1 = 7$, which is odd.
when $L = 8$, $2L + 1 = 17$, which is odd.
when $L = 13$, $2L + 1 = 27$, which is odd.
when $L = 50$, $2L + 1 = 101$, which is odd.

And again, from logical reasoning we can show that the number $2L + 1$ must be odd. The explanation lies in the fact that the number $2L + 1$ is 1 more than the number $2L$, which itself must be even because it contains the factor 2. A number one greater than an even number is necessarily odd.

With these ways of representing even and odd numbers at our disposal, we are now ready to prove the earlier result algebraically. I have restated it below in Example 4.

EXAMPLE 4

Prove algebraically the result that the sum of two odd numbers always gives an even number.

Solution

One possibility might be to let both the two odd numbers be represented by $2K + 1$. However, the problem with doing this is that, whatever value for K is chosen, we find ourselves with two odd numbers with the same value. This is subtly different from the problem we set out to prove. We need to allow the two odd numbers to be different, so, using different letters, we can let them be $2K + 1$ and $2L + 1$, respectively.

Their sum $= (2K + 1) + (2L + 1)$

Simplifying, we get $2K + 2L + 1 + 1 = 2K + 2L + 2$

Now, notice that $2K + 2L + 2$ can be written as $2(K + L + 1)$.

Since this number contains a factor of 2, and $K + L + 1$ is a whole number, then $2(K + L + 1)$ must be even.

We have now proved the result *in general terms.* No matter what whole number values you think up for K and L, the general argument demonstrates that the result $2(K + L + 1)$ will always be an even number.

If you are unfamiliar with algebraic reasoning, you may need to read this proof through more than once. Then, when you are more confident, have a try at writing your own proof in your answer to Exercise 9.6.

By the way, don't worry if you find this difficult. Most people find algebraic proofs hard to fathom and you would need a lot more practice at working with algebraic symbols than has been provided in this chapter if you are to perform proofs with confidence. I have included it mostly to indicate the sort of things that mathematicians spend their time on, and to give you an insight into how algebraic symbols can be an aid in solving abstract problems.

EXERCISE 9.6 Proving with algebra

Is it true that the product of two odd numbers is always an odd number? (*Reminder*: product means 'the result of multiplying together'.)

Test this out first with a few special cases and then try to prove it algebraically.

Spreadsheets

	A	B	C	D	E
1					
2					
3		Cell B3			

A spreadsheet is a computer tool that is used to set out data in rows and columns on a screen. The rows are numbered 1, 2, 3 … down the left-hand side, while the columns are labelled A, B, C … across the top. A typical spreadsheet might resemble the table shown above, except that more rows and columns are visible on the screen at any one time.

Each 'cell' is a location where information can be stored. Cells are identified by their column and row position. For example, cell B3 is indicated in the table shown above; it is the cell in row 3 of column B.

The information you might want to put into each cell will be one of three basic types:

- **Numbers.** These can be either whole numbers or decimals, for example 7, 120, 6.32.
- **Words.** These can either be headings or explanatory text.
- **Formulae.** The real power of a spreadsheet is its ability to handle formulae.

When you enter a suitable formula into a particular cell, this formula will contain one or more references to other cells. The cell will then display a value calculated from the numbers currently stored in the cells referred to in this formula; this could be an average or perhaps a row or column total. If these cell values are altered, the formula will instantly recalculate on the basis of the updated values and display the new result. To take a simple example, suppose I entered my height in metres (1.72 m) into cell A1. I could then enter a formula into cell B1 for converting the height into centimetres (type = A1*100). Press the <ENTER> key to 'enter' the formula and the value 172 is displayed in B2. Now enter any new height in metres into cell A1 and the value in B2 adjusts to display the height in centimetres.

A spreadsheet is useful for storing and processing data when repeated calculations of a similar nature are required. Next to word processing, a spreadsheet is the most frequently used tool in business. It is also extremely useful for householders to help solve problems that crop up in their various life roles as consumers, tax payers, members of community organizations, etc. For example, it can be used to investigate questions such as:

- How much will this journey cost for different groups of people?
- Is my bank statement correct?
- Which of these buys offers the best value for money?
- What is the calorie count of these various meals?
- What would these values look like sorted in order from smallest to biggest?
- How can I quickly express all these figures as percentages?

A spreadsheet is a powerful tool for carrying out repeated calculations. You simply perform the first calculation and then a further command will complete all the other calculations automatically. Another advantage of a spreadsheet over pencil and paper is its size. The grid that appears on the screen is actually only a window on a much larger grid. In fact, most spreadsheets have hundreds of rows and columns, should you need to use them. Movement around the spreadsheet is also straightforward – you can use certain keys to move to adjacent cells or to another cell of your choice.

Once the data has been entered into the spreadsheet, there are a variety of options available for helping to make better sense of the figures. For example, columns or rows can be re-ordered or sorted either alphabetically or according to size. Row and column totals can be quickly found and entire rows or columns can be converted into percentage form. A variety of summary values can be calculated (mean, mode, range and so on). Finally, spreadsheets contain powerful graphing facilities that enable you to display some or all of the data as a piechart, bargraph, scattergraph and so on.

Summary

A key point made in the introduction to this chapter was that it is silly to criticize algebra because it is abstract. Essentially, the purpose of algebra *is* to be abstract. Algebra involves expressing relationships in a mathematical shorthand in the form of symbols and letters. This has the effect of reducing the problem to its bare essentials and allows you to see and manipulate its main features. Certain algebraic conventions were explained (for example, that writing $4X$ actually means '4 times X').

As the title suggests, the main activities of the chapter involved using formulas and the next section invited you to dip your toe into the esoteric world of mathematical proof. The chapter ended with a look at a most useful piece of computer software – a spreadsheet.

Answers to exercises for Chapter 09

9.1

Example	Source	Meaning
a det. hse, lge gdns, gd decs, FCH ...	Newspaper ad. for a house	Detached house, large gardens, good decorations, full central heating ...
b K1, P1, M1, C4F, K2, ...	Knitting	Knit 1, Purl 1, Make 1, Cable 4 forward, Knit 2
c PAS, MOT, fsh	Newspaper ad. for a car	Power assisted steering, holds an MOT certificate, full service history
d NYWJM seeks same with view to B&D, S&M, etc.	Personal ad. in a US newspaper	New York white Jewish male . . . Bondage & Discipline, Sadism & Masochism, etc.

9.2

Longhand	Shorthand
a Three multiplied by two and a half	$3 \times 2\frac{1}{2}$
b The sum of twelve, and four and three quarters	$12 + 4\frac{3}{4}$
c The sum of the squares of three and four	$3^2 + 4^2$
d Four times the difference of nine and three	$4(9 - 3)$
e Five plus four, all divided by the product of five and four	$\frac{5+4}{5 \times 4}$
f The number of inches, I, is found by multiplying the number of metres, M, by 39.37.	$I = 39.37 \times M$

9.3 a A ten-year-old's dosage = $24 \times \frac{10}{10+12} = 24 \times \frac{10}{22} = 10.9\mu g$.

   b A four-year-old's dosage = $200 \times \frac{4}{4+12} = 200 \times \frac{4}{16} = 50\mu g$.

9.4 a The temperature in degrees Fahrenheit, $F = 1.8 \times 30 + 32$
$$= 86°F.$$

   b The temperature in degrees Fahrenheit, $F = 1.8 \times 100 + 32$
$$= 212°F.$$

   c You might have tried to find this temperature by trial and error. The solution is −40. This can be checked by putting the value −40°C into the formula, and the result −40°F comes out.
Thus:
The temperature in degrees Fahrenheit, $F = 1.8 \times -40 + 32$
$$= -40°F$$

The answer can be calculated directly using algebra, as follows. Let the unknown temperature = T.
If the temperature $T°F = T°C$, then they are connected by the formula, as follows.
$$T = 1.8T + 32$$
The table below summarizes how this equation can now be solved.

Algebra	Explanation
$T = 1.8T + 32$	This is the equation to be solved
$T - 1.8T = 1.8T - 1.8T + 32$	Subtract $1.8T$ from both sides (1)
$-0.8T = 32$	Simplify the terms in T
$\frac{-0.8T}{-0.8} = \frac{32}{-0.8}$	Divide both sides by -0.8 (2)
$T = \frac{32}{-0.8}$	Simplify
$= -40$	The solution of the equation is -40

   1 The intention here is to collect the T terms on one side of the = and leave the number on the other side.

   2 The intention here is to isolate the T on its own. Remember that the object of the exercise is to find the value of T.

9.5 The formula is: $C = 20.16 + 0.042U$

   a $C = 20.16 + (0.042 \times 944) = £59.81$ (rounded to the nearest penny).

   b $C = 20.16 + (0.042 \times 3122) = £151.28$ (rounded to the nearest penny).

9.6 Proving with algebra

It is true that the product of two odd numbers is always an odd number.

First here are some special cases:

$3 \times 5 = 15$, which is odd

$5 \times 9 = 45$, which is odd

$13 \times 7 = 91$, which is odd

$113 \times 5613 = 634\,269$, which is odd.

Now we move on to a general algebraic solution.

As before, we can let these two odd numbers be represented by $2K + 1$ and $2L + 1$.

Their product $= (2K + 1) \times (2L + 1)$

This can be written as $2K(2L + 1) + 1(2L + 1)$

implifying, we get $\quad 4KL + 2K + 2L + 1$

Ignoring the final term, the '1' for the moment, notice that the first three terms, $4KL + 2K + 2L$ can be written as $2(2KL + K + L)$.

Since this number contains a factor of 2, and $2KL + K + L$ is a whole number, then $4KL + 2K + 2L$ must be even.

Now add the final '1' and it follows that $4KL + 2K + 2L + 1$ is odd.

10

puzzles, games and diversions

This chapter provides a few suggestions for number puzzles and activities which should help to amuse and entertain on a long journey or a wet weekend.

1 Number plate games

Games with car number plates can be played anywhere near a road or car park or on a long journey. See if you can spot numbers where:

(i) the digits add to 10 (e.g. T163 VVV)
(ii) all the digits are even (e.g. S426 EJS)
(iii) all the digits are odd (e.g. X715 FLO)
(iv) all the digits are prime numbers (e.g. T725 WXC)

With practice, most people quickly get good at looking for and spotting patterns in numbers, in which case the game can be made more challenging. For example, try to spot numbers which are the product of two primes (e.g. W247 TTT: 247 is the product of 13 and 19, which are both prime).

The renowned Indian mathematician, Srinivasa Ramanujan was once visited by the British mathematician G. H. Hardy. Hardy remarked that he had just travelled in a taxi bearing the rather dull number 1729.

'On the contrary', said his friend, 'it is a very interesting number. It is the smallest number expressible as a sum of two cubes in two different ways' (10 cubed plus 9 cubed or 12 cubed plus 1 cubed).

Not everyone has quite the same fascination and skill with number properties as Ramanujan. But you should not under-estimate the degree of interest and social cachet you are likely to attract at dinner parties by passing on gems about the properties of certain numbers. For example, a perfect opening line during that awkward 'first introduced' phase at a party might be, 'Did you know that our host's telephone number is the first six digits of the decimal expansion of pi, backwards?'

Well, this sort of chat-up line certainly seems to work for me!

---

2 Pub cricket

This is played on a car or coach journey where the route is likely to pass a number of pubs.

Players take turns to 'bat'. You score according to the number of legs (human or animal) which can be seen on each pub sign that you drive past. For example, the 'Bull and Butcher' scores six (four for the bull and two for the butcher), while the 'White Hart' scores four. If you go past a pub sign which has no legs, then you are 'out' and the next player takes a turn at batting.

Popular signs in this game, by the way, are the 'Coach and Horses' (with 24 or more legs, depending on the number of horses) and the 'Cricketers' Arms' (with up to thirty, depending on whether or not both batsmen and both umpires are depicted!)

3 Guess my number

One player picks a number between 1 and 100 and the other player must guess it with as few questions as possible. Note that the questions must be such that they require a Yes/No answer.
Useful questions are ones such as:
'Is it less than 50?' or 'Is it even?'
Less useful questions are ones such as:
'Is it 26?' since this eliminates only one number at a time.

4 The story of 12

How many *different* stories can you make of 12?
Well, it is $11 + 1$, $10 + 2$, etc.
Don't forget it is also $12 + 0$.
What about 4×3, 3×4 and 6×2?
And don't forget 12×1. Now we're getting stuck.
Ah! it's 24 lots of a half, so it's $24 \times \frac{1}{2}$.
Well, if you're allowing fractions it's $1\frac{3}{4} + 10\frac{1}{4}$.
Hey, this could go on all night!

5 Finger tables

Most people know their 'times table' up to about 5. However, with 6 and above they may have problems. Here is a method which provides the answer to all products between 6 and 10 (remember that in mathematics 'product' means what you get when you multiply), using the cheapest digital calculator around – your fingers.

Place your hands in front of you as in the diagram, thumbs uppermost.

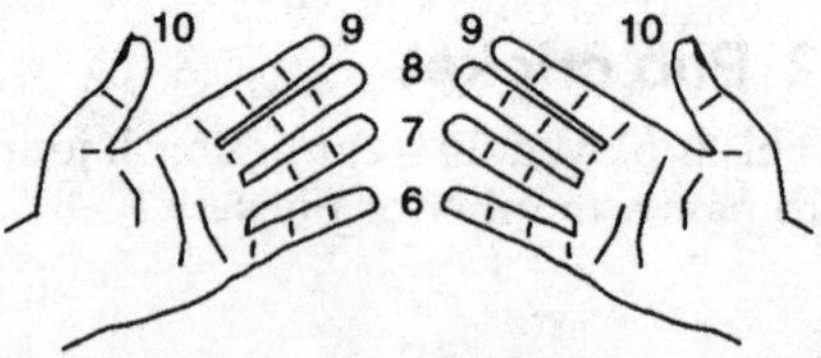

Starting with the thumb, number the fingers as shown. Now to multiply, say, 7 and 8, touch the 7 finger of one hand with the 8 finger of the other (it doesn't matter which way round).

Now the answer to 8 × 7 can be found as follows:
a Count the number of fingers *below and including* the touching fingers (in this case 5). This gives the number of tens in the answer.
b Multiply the number of fingers on each hand *above* the touching fingers (here it is 3 × 2 = 6). This gives the number of units in the answer.

So the answer is fifty-six.

Try it for some other numbers.
Why does it always work?
You will need to do some algebra to prove it works every time!

6 Magic squares

This is a magic square.

8	1	6
3	5	7
4	9	2

A magic square is so named because all the rows, columns and diagonals 'magically' add to the same total (in this case, 15). This is known as a 3 by 3 magic square because there are three rows and three columns.

• Can you make a different 3 by 3 magic square so that all the rows, columns and diagonals add to 15?
• How about a 4 by 4 magic square … or a 5 by 5?

Hint: You may have spotted that the total of 15 for the 3 by 3 magic square is three times five (5 is both the number in the centre square and the middle value in the range of 1 to 9). Can you first of all work out what the rows, columns and diagonals of the 4 by 4 square should add to?

7 Magic triangle

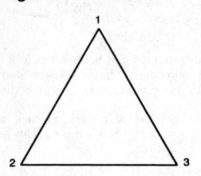

The numbers 1, 2 and 3 have been placed at the vertices (i.e. the corners) of the triangle.

Can you place two of the following numbers 4, 5, 6, 7, 8 and 9 along each side of the triangle so that the four numbers on each side (i.e. including the numbers at the vertices) add up to 17?

8 Upside down

The year 1961 reads the same when turned upside down.
When was the most recent year prior to 1961 that reads the same upside down?

When will be the next year that this works?

Explore what happens when letters and number are turned upside down. For example, what digits become letters of the alphabet when turned upside down?

9 Logically speaking

a What word is from this sentence?
b 1981 is to 1861 as 8901 is to what?
c Seven numbers can be seen in this sentence, but you know three of them are written out backwards.
What are they?

d Ten birds sit on a roof. You make a noise to scare them off, and all but four of them fly away. How many are left?

e How old is a coin engraved with the date 88BC?

f You have three pairs of different-coloured socks in a drawer, each sock separated from its partner. How many single socks do you take out, without looking, to be sure that you have:
(i) a matching pair of any colour?
(ii) a matching pair of a particular colour?

g You need seven candle stubs to make a new candle. How many candles will you be able to make if you start with 49 stubs?

h What is the fewest number of coins you need to pay for something costing 85p without receiving change?

i Fill in the missing signs in the sums below.

$$2 \boxed{} 4 \boxed{} 3 = 5$$
$$2 \boxed{} 2 \boxed{} 2 = 3$$
$$8 \boxed{} 2 \boxed{} 2 = 2$$

10 Calculator inversions

Enter the number 53045 on your calculator. Now turn it upside down. Can you see something that helps you keep your feet off the ground?

Next, try decoding the following 'message':

$3145 \times 10 + 123$
$204 \div 4$
$0.5 \times 0.5 \times 2$
257×3

Finally, have a go at the following crossword puzzle. Use the results of each calculation, upside down, to fit the puzzle.

Across

1 $3 \times 1000 + 45$ may be a good fit (4)

6 $5 \times 1111 - 18$ is no more (4)

7 $9^3 - 19$ is useful for troubled waters (3)

8 $2^5 + 11$ but please speak up! (2)

10 123×25 for a quick gin? (4)

12 Two score, for a surprise (2)

14 $13 \times 24439 + 5000000$, so stop and buy some (7)

16 $\frac{432}{6} + 1$ for the Spanish (2)

17 $\frac{70}{100}$ Verily, it sounds like a cow hath spoken (2)

Down

2 $0.65 + 0.1234$ is one third of a police officer's greeting (5)

3 $17 \times 10 \times 7 \times 3 + 3$ is another possibility (4)

4 $11^2 \times 31$ is emerald in Ireland (4)

5 $1111 \times 5 - 48$ for a no-win situation (4)

9 Two fifths expressed as a decimal is one third of a Christmas greeting (2)

11 1101×7 describes life in the 10 across lane? (4)

13 $19 \times 2 \times 193$ can be found on a 1 across (4)

15 Half could be a needle pulling thread (2)

11 Four 4s

The number 7 can be expressed using four 4s as follows:

$$7 = 4 + 4 - \frac{4}{4}$$

Express the numbers 0, 1, 2 ... 10 using exactly four 4s and any other operation (for example, +, −, ×, ÷). You are also allowed square root, $\sqrt{\ }$.

Hint: 2 can be written as $\frac{4+4}{4}$.

12 The bells, the bells!

a If it takes 15 seconds for a church bell to chime 6 o'clock, how long does it take to chime midnight? (It's not as easy as it sounds!)
b A fence panel is 2 m long. How many fence posts are needed to panel a 16 m gap?

13 Return journey

I plan to complete a return journey (there and back) in an average time of 40 mph. However, my outward journey is slow and I complete that part at 20 mph. How fast must I travel on the return journey to average 40 mph overall?

14 Find the numbers

a Two consecutive numbers add to give 49. What are the numbers?
b Three consecutive numbers have a total of 60. What are the numbers?
c Two consecutive numbers have a product of 600. What are the numbers? (*Note*: The product is what you get when you multiply.)
d Three consecutive numbers have a product of 1716. What are the numbers?
e Two numbers have a difference of 15 and a product of 54. What are the numbers?

15 Explore and explain the pattern

Try these out on your calculator
a 37×3, 37×33, 37×333, etc.
b 1^2, 11^2, 111^2, 1111^2, ...

c Take three consecutive numbers   (say, 8, 9 and 10)
  Multiply the first and third number   ($8 \times 10 =$)
  Square the middle number   ($9 \times 9 =$)
  Subtract smaller answer from bigger answer
  The result is 1   ($81 - 80 = 1$)

Does this always work for three consecutive numbers?

16 Large and small sums

Try to arrange the digits 1, 2, 3, 4 and 5 (each used once only) to form two numbers so that the sum of the numbers is as large as possible.

Now try to arrange them so their sum is as small as possible.

Explore the same sorts of questions for multiplication, ...

17 1089 and all that

Take a three-digit number	say, 724
Reverse the digits and subtract whichever is the smaller from the bigger.	$724 - 427 = 297$
Reverse the digits of the answer and add it to the answer	$792 + 297 = 1089$

Try it for other three-digit numbers. Why does 1089 keep cropping up?

Do you always get 1089?

If not, then which numbers does it not work for. Why?

> By the way, this puzzle can be set up as a trick to impress your friends, as follows.
> Write down the figure 1089 on a piece of paper and seal it in an envelope beforehand.
> Then ask the friend to choose any three-digit number and perform the calculation described above.
> *Note*: If a zero occurs in any part of the calculation, this must be counted as well.
> For example:
>
> $534 - 435 = 099$

Reversed, 099 becomes 990. This gives 099 + 990 = 1089.
In order to give the trick a little 'pzazz', ask your 'victim' for some additional but totally irrelevant information (for example, date of birth, telephone number, favourite colour, and so on).

18 Gold pieces

In exchange for your lucky calculator and this oh-so-precious maths textbook, the evil, cunning and extremely wealthy empress Calcula offers you a choice of one of the following options.

You will be given:
either
a As many gold pieces as the number of minutes you have been alive;
or
b As many gold pieces as the largest number you can get on your calculator by pressing just five keys;
or
c One gold piece on the first day of this month, two on the second, four on the third, eight on the fourth, and so on, ending on the last day of the month.

With the help of your calculator, decide what you should do.

19 Initially speaking

This puzzle is easiest to explain with the following example.

Clue   *Solution*
3 B M, S H T R   3 Blind Mice, See How They Run

Now complete the solutions below.
If you get them all right, it should spell out the name of a film as well as a catch phrase when you read down the central boxes.

3 BLIND MICE [S] EE HOW THEY RUN

101 _____ [ ] __

3 [ ] __ [K] _____

10 [Y] ____ [I][N] [A] _ [ ] ____

18 [H] ____ [I][N] [A] __ [ ] _ _____

24 [B] _____ [ ] ___ [B] ____ [I][N] [A] ___

WE 3 [ ] ____ [F] ___ [O] _____ ___

THE 10 [C] _____ [ ] ___

36 [ ] _____ [I][N] [A][Y] ___

76 [T] _____ [I][N] [ ] __ ___ _____

64 [S] _____ [O][N] [A] _ [ ] ___ _____

11 [P] _____ [I][N] [A] __ [ ] _____ [T] ___

100 I [S] ___ _____ ____ [ ] [O][F] [W] ____

20 Nim

Nim is one of the oldest recorded games, possibly Chinese in origin, and is usually played by two people. There are many versions of Nim, one of which is described below.

Start with a pile of matchsticks. Each player, in turn, removes at least one but not more than six matches. The winner is the player who picks up the last match.

Here is a typical game.
Start with 28 matches in the pile.
A picks up 4 matches, leaving 24
B picks up 5 matches, leaving 19
A picks up 2 matches, leaving 17
B picks up 6 matches, leaving 11
A picks up 3 matches, leaving 8
B picks up 1 match, leaving 7
A (realizing that defeat is just a match away) picks up 1 match, leaving 6
B picks up all remaining 6 matches, thereby winning game, set … and match.

21 Guess the number

The rules of this game are given at the end of Chapter 05.

22 Calculator snooker

Player A enters any two-digit number. *B* takes a 'shot' by performing a multiplication sum. To 'pot' a ball, the first digit of the answer must be correct according to the table shown. (The degree of accuracy can be varied according to experience.)

Ball	Red	Yellow	Green	Brown	Blue	Pink	Black
Result needed	1...	2...	3...	4...	5...	6...	7...
Score	1	2	3	4	5	6	7

Otherwise, the rules are similar to 'real' snooker. There are 10 (or 15) reds and one of each of the six 'colours'. A player must score in the order red, colour, red, colour, and so on, until all the reds have gone. (*Note*: The colours are replaced but the reds are not.) When the last red has gone, the colours are potted 'in order' and are not replaced.

For example, one sequence of plays was:

Player	Enters	Display	Comments
Jimmy	69	69	
Peta	× 2 =	138	Peta pots the first red.
			She elects to go for blue ...
	× 5 =	690	... and misses.
Karen	× 2 =	1380	Karen pots the second red.
			She elects to go for black ...
	× 5.5 =	7590	... and pots it.
	× 1.6 =	12144	The third red.
			She elects to go for black again ...
	× 7 =	85008	... and misses.

23 Place invaders *(a game for one or two players)*

This game can be played at different levels (1, 2, 3, etc.). Move on to a new level when you find the game too easy.

Place invaders 1

Enter a 3-digit number into the calculator (say 352). These three digits are removed one at a time by subtracting to zero.
Example: Starting number *352*

Key presses	Display
$-$ 2 $=$	350
$-$ 50 $=$	300
$-$ 300 $=$	0

You can make up your own numbers and let your partner remove them

Try the following: 416, 143, 385, 512, 853, 264, 179, 954, 589, 741.

Place invaders 2

The same as Place invaders 1 except that the digits must be removed in ascending order.

Example: Starting number <u>352</u>

Key presses	Display
$-$ 2 $=$	350
$-$ 300 $=$	50
$-$ 50 $=$	0

i.e. you remove the 2, then the 3, and then the 5

Place invaders 3

The same as Place invaders 2, except that you can use numbers with more digits. Try 4-digit numbers, then 5, 6, 7 and 8.

Place invaders 4

The same as Place invaders 3, except that you remove the digits by addition, not subtraction. This time the game will end with a 1 followed by a string of zeros. Use as few goes as possible.

Example: Starting number <u>1736</u>

Key presses	Display
$+$ 4 $=$	1740
$+$ 60 $=$	1800
$+$ 200 $=$	2000
$+$ 8000 $=$	10000

Note: If you start with a 5-digit number, the game ends with a display of 100000.

Place invaders 5

The same as Place invaders 3, except that you can use decimals, e.g. 451.326 to be removed by subtraction in the order of 1, 2, 3, 4, 5, 6.

Note: If you make a mistake, it is easy to undo by adding back the number that you have just subtracted.

Answers for Chapter 10

1–5
No comments.

6
Here is another 3 by 3 magic square

8	1	6
3	5	7
4	9	2

Now here is a 4 by 4 magic square.

16	3	2	13
5	10	11	8
9	6	7	12
4	15	14	1

With this 4 by 4 magic square each row, column and diagonal adds to 34. What makes this one even more magic is that each block of four corner squares also adds to 34.

Hang on – what about the four central numbers ...?

7

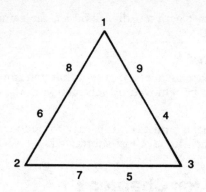

8 a 1881   **b** 8008

9 a What word is *missing* from this sentence?
b 1981 is to 1861 as 8901 is to *8601*.
c **Seven** numbers can be se**en in** this sentence, but you kn**ow three** of them are writt**en o**ut backwards.
d Four birds are left.
e No coin could have been engraved with this date.
f (i) 4 socks   (ii) 6 socks
g Seven candles initially. But these will produce seven more butts, so the answer is eight.
h Four coins: 50p + 20p + 10p + 5p
i 2 ⊞ 4 ⊟ 3 = 5
  2 ⊡ 2 ⊞ 2 = 3
  8 ⊡ 2 ⊟ 2 = 2   or 8 ⊡ 2 ⊡ 2 = 2

10 ELSiE iS SO iLL

11
$0 = 4 + 4 - 4 - 4$
$1 = \frac{4+4}{4+4}$ or $\frac{44}{44}$ or $\frac{4}{4} + 4 - 4$
$2 = \frac{4}{4} + \frac{4}{4}$
$3 = \frac{4+4+4}{4}$
$4 = 4 + \frac{4-4}{4}$
$5 = \sqrt{4} + \sqrt{4} + \frac{4}{4}$
$6 = 4 + \frac{4+4}{4}$
$7 = \frac{44}{4} - 4$ or $4 + 4 - \frac{4}{4}$
$8 = 4 + 4 + 4 - 4$
$9 = 4 + 4 + \frac{4}{4}$
$10 = 4 + 4 + \frac{4}{\sqrt{4}}$

12 a Answer 33 seconds.
It takes 15 seconds for 6 chimes. There are 5 intervals between the first and the sixth chime. Therefore it must take 3 seconds per interval. Twelve chimes has 11 intervals, hence 33 seconds.

b 9 posts are needed for 8 spaces.
Both the above questions refer to a common 'type' of maths question known as the old 'posts and spaces' trick. There are two things to remember in any 'posts and spaces' sort of situation. Firstly, be clear about which you are trying to count, the 'posts' or the 'spaces'. And secondly, remember that there is always one fewer 'space' than there are 'posts'.

13 It can't be done! Suppose the total distance (there and back) is 40 miles, then the total journey (there and back) must take exactly one hour. If the outward journey of 20 miles is completed at a speed of 20 mph, the one hour is completely used up!

14 a 24 and 25.
b 19, 20 and 21.
c 24 and 25.
d 11, 12 and 13.
e 3 and 18.

15 a 111, 1221, 12321. These numbers are palindromes (i.e. they read the same backwards as forwards).
b 1, 121, 12321, 1234321. Again a palindromic sequence similar, but not identical to part **a**.
c This result works for all sets of three consecutive numbers.

16 The largest sum is 573 (541 + 32)
The smallest sum is 159 (125 + 34)
The largest product is 22 403 (521 × 43)
The smallest product is 3185 (245 × 13).

17 This result works for most but not all numbers. Try starting with some palindromic numbers and see what happens.

18 Let us calculate each option in turn.

   a As many gold pieces as the number of minutes you have been alive.

     Assuming that you are, say, forty-five years old, the required calculation is:

     $45 \times 365 \times 24 \times 60 = 23\,652\,000$.

     In other words, between 23 million and 24 million.

   b As many gold pieces as the largest number you can get on your calculator by pressing just five keys. My best effort here was to press:

     99 $\boxed{\times}$ 9 $\boxed{=}$

     This comes up with a puny 891 gold pieces. If your calculator has a 'square' function, marked $\boxed{x^2}$ you will do much better than this. So much so that even five key presses will produce an answer too large for most calculators to display and which will therefore result in an error message. For example:

     99 $\boxed{x^2}$ $\boxed{x^2}$ $\boxed{x^2}$ produces an error message on my calculator.

   c One gold piece on the first day of this month, two on the second, four on the third, eight on the fourth, and so on, ending on the last day of the month.

     This arrangement may sound very low key, but in fact it will produce an astronomically large result quite quickly. The best way to get an impression of the effects of doubling is to set your calculator's constant to multiply by 2 and then keep pressing $\boxed{=}$. What you will see should be something like the following:

Day	1	2	3	4	5	6	... 12	... 16
Amount	1	2	4	8	16	32	... 2048	... 32 768

     Before you get to the end of the month you will probably find that the calculator has over-stretched itself and produced an error message!

     The answer, therefore, is that the 'best' option to choose depends on what sort of features your calculator has – for example, how many figures it displays, which keys it provides and so on. But whichever calculator you use, the third option is certainly a good one to go for!

19 The solution is 'Some like it hot'.

3 BLIND MICE $\boxed{S}$ EE HOW THEY RUN

101 DALMATI $\boxed{O}$ NS

3 $\boxed{M}$ US $\boxed{K}$ ATEERS

10 $\boxed{Y}$ EARS $\boxed{I}\boxed{N}$ $\boxed{A}$ D $\boxed{E}$ CADE

18 $\boxed{H}$ OLES $\boxed{I}\boxed{N}$ $\boxed{A}$ GO $\boxed{L}$ F COURSE

24 $\boxed{B}$ LACKB $\boxed{I}$ RDS $\boxed{B}$ AKED $\boxed{I}\boxed{N}$ $\boxed{A}$ PIE

WE 3 $\boxed{K}$ INGS $\boxed{F}$ ROM $\boxed{O}$ RIENT ARE

THE 10 $\boxed{C}$ OMMANDM $\boxed{E}$ NTS

36 $\boxed{I}$ INCHES $\boxed{I}\boxed{N}$ $\boxed{A}$ $\boxed{Y}$ ARD

76 $\boxed{T}$ ROMBONES $\boxed{I}\boxed{N}$ $\boxed{T}$ HE BIG PARADE

64 $\boxed{S}$ QUARES $\boxed{O}\boxed{N}$ $\boxed{A}$ C $\boxed{H}$ ESS BOARD

11 $\boxed{P}$ LAYERS $\boxed{I}\boxed{N}$ $\boxed{A}$ FO $\boxed{O}$ TBALL $\boxed{T}$ EAM

100 $\boxed{I}\boxed{S}$ THE BOILING POIN $\boxed{T}$ $\boxed{O}$ F $\boxed{W}$ ATER

20–23 No comments.

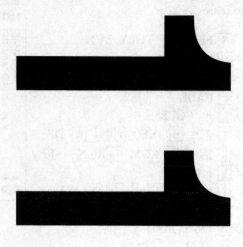

diagnostic quiz

In this chapter you will learn:
- how much maths you have already learnt by reading this book!

Now that you have carefully read through every page of the preceding ten chapters (well, maybe you skipped a few pages!), you might like to take stock of what you have learnt by trying to answer the questions in this diagnostic quiz.

A quiz, or test, can be tackled in many different ways. If you get every question right, you may feel good about yourself, but you probably haven't learnt anything from it. If you get all the questions wrong, you will probably feel pretty depressed and unable to exploit the learning opportunities offered by the experience. I hope that you will be somewhere in between. This quiz is not designed to trick you or to make you feel depressed. Having said that, you are very unlikely to find all the questions easy or to get every question right.

Here are some guidelines for tackling the quiz.

- You should be prepared to use your calculator for every question, except for Question 1 where you are asked not to.
- Read each question carefully before you do it, so that you *answer exactly the question that has been asked*. For example, if it asks you to write numbers in order from smallest to largest, don't give your answers from largest to smallest.
- Don't be afraid to look things up in earlier chapters of the book if you have forgotten, say, how to convert miles into kilometres. This isn't a test to be taken under examination conditions and you aren't expected to remember all the formulas and conversions in your head.

So, please give the quiz your very best shot. It is designed to take about one hour, but be prepared to take longer than that if you need to. When you have done all that you can do, then work through my solutions at the end of the chapter. As you will see, I have included detailed comments after the solutions in order that you can 'turn your errors into learning opportunities'. For each one that you answered incorrectly, ask yourself the following questions:

- 'WHERE AND WHY HAVE I GONE WRONG?'
- 'WHAT CAN I LEARN FROM THIS?'

Good luck!

Quiz

1 Try these calculations without using your calculator.
   a (i) 3×14; (ii) $-5 - 17$; (iii) $-20 \div 4$; (iv) $15 \div 75$.
   b (i) $2\frac{1}{2} + 1\frac{1}{4}$; (ii) $3\frac{1}{2} - 2\frac{3}{4}$; (iii) $5\frac{1}{2} \times 3$; (iv) $3\frac{2}{3} \times 5$
   c (i) 8^2; (ii) $\sqrt{81}$; (iii) $\sqrt{(3^2 + 4^2)}$

2 Express the following as decimal numbers.
   a $20 + 7 + \frac{3}{10} + \frac{6}{100} + \frac{4}{1000}$
   b $60 - 3 + \frac{8}{10} + \frac{2}{100} - \frac{7}{1000}$
   c $\frac{62341}{1000}$

3 In the number 13.873, the '7' represents the number of hundredths.
   What does the '7' represent in the following numbers?
   a 271.93
   b 11.724
   c 0.00117

4 Which is bigger:
   a three quarters or 70 per cent?
   b 0.06 or one twentieth?
   c two fifths or 0.5?
   d 10 per cent or an eighth?
   e 8 per cent or a tenth?

5 Using suitable *metric* units, estimate the following:
   a the height of a chair seat from the floor
   b the width of a cooker
   c the weight of a new-born baby
   d the distance from London to Birmingham
   e the capacity of a doorstep milk bottle
   f the weight of a letter
   g the temperature inside a domestic refrigerator
   h two teaspoonfuls of liquid
   i the thickness of a £1 coin
   j the temperature on a hot summer's day in London.

6 Place the following in order of size *from smallest to largest*
   a 450ml; half a litre; one pint; 75 centilitres
   b $1\frac{3}{4}$ metres; 18 cm; 1.2 km; 300 mm
   c half a week; four days; 95 hours; 0.01 of a year.

7 *Using both imperial and metric units*, estimate the following:

   a the speed of a car travelling on the outside lane of a motorway in the UK

   b the speed of someone having a brisk walk

   c the speed of a top 100 m runner

   d the speed of a supersonic jet aircraft.

8 'It is possible to fit the world's population on the Isle of Wight.'

Use the following facts to check this claim.

- The Isle of Wight has an area of 381 km^2.
- It is possible to squeeze about 10 average-sized people into one m^2.
- The world's population, at the time of writing, is roughly 6 billion.

9 The table below lists, in thousands of pounds (£000), the voluntary cash donations to the top 10 UK charities in a particular year.

Name	Income (£000)
National Trust	78 745
Oxfam	58 972
RNLI	56 229
Save the Children Fund	53 866
Imperial Cancer Research Fund	48 395
Cancer Research Campaign	45 352
Barnardos	36 452
Help the Aged	33 141
Salvation Army	32 303
NSPCC	30 818

   a Rewrite the ten incomes, rounded to the nearest £million.

   b Using the rounded figures, sketch a horizontal barchart to represent the earnings of the top five charities in a particular year.

   c Explain why a piechart would not be appropriate for depicting the earnings of the top five charities.

10 In a certain year, an estimated 19.2 million international visitors came to Britain and spent £9.2 billion. The piechart below shows the estimated annual tourist spending, broken down by which part of the world the tourists came from.

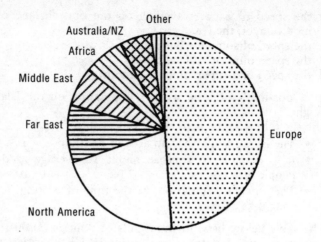

a From the piechart, estimate the annual spending by tourists from Europe.

b From which one of the regions listed here did roughly £2 billion of the tourist revenue come?

c If Britain were able to attract an extra 10 million visitors, how much more revenue might it be able to generate? Make a note of any assumptions that you have made in doing this calculation.

11 A survey of toy prices was taken in three large stores. The prices (in £) of three toys are summarized below.

	Hamleys	John Lewis	Toys Я Us
Toy A	12.99	9.25	9.29
Toy B	29.00	22.75	22.99
Toy C	7.99	5.95	4.97

On the basis of these prices:

a which of the three stores seems to be the most expensive?

b which store is the cheapest?

c If you bought all three items at the cheapest price on offer, how much would you have saved compared with buying them at the most expensive price?

d From your answer to part **c**, calculate your total savings as a percentage of the total cheapest price.

12 Petra earns £6200 per year doing part-time work. She pays tax at the basic rate of 20% and her tax allowances are £3750.

The annual amount that she has to pay in Tax, T, can be calculated from the following formula.

$$T = 0.2 (I - A)$$

(where I represents her annual income and A represents her tax allowances)

a Explain in your own words how, according to the tax formula, annual tax on earnings is calculated.

b Calculate how much Petra will have to pay in tax:
  (i) over the year
  (ii) each week. (Give your answer to the nearest penny.)

c Assuming there are no other stoppages from her wages, calculate how much Petra will receive each week after the weekly tax bill has been paid. (Give your answer to the nearest penny.)

Solutions to the quiz

1 a (i) 42; (ii) –22; (iii) –5; (iv) $\frac{1}{5}$ or 0.2
   b (i) $3\frac{3}{4}$; (ii) $\frac{3}{4}$; (iii) $16\frac{1}{2}$; (iv) $18\frac{1}{3}$
   c (i) 64; (ii) 9 (or –9); (iii) 5 (or –5)

2 a 27.364
   b 57.813
   c 62.341

3 a tens
   b tenths
   c hundred thousandths

4 a three quarters
   b 0.06
   c 0.5
   d an eighth
   e a tenth

5 For this question, you must get *both* the correct answer *and* the correct units.
   a about 45 cm (I will accept answers between 40 and 50 cm)
   b about 60 cm, or 600 mm (I will accept answers between 50 and 70 cm)
   c about 4 kg, or 4000 g (I will accept answers between 3 and 5 kg)
   d about 175 km (I will accept answers between 150 and 200 km)
   e about 570 ml (I will accept answers between 500 and 600 ml)
   f about 40 g (I will accept answers between 10 and 60 g)
   g between 0°C and 5°C

h about 12 ml (I will accept answers between 10 and 15 ml)
i about 3 mm (I will accept answers between 2 and 4 mm)
j about 30°C (I will accept answers between 25 and 35°C)

6 a 450 ml, half a litre, one pint, 75 centilitres
b 18 cm, 300 mm, $1\frac{3}{4}$ m, 1.2 km
c half a week, 0.01 of a year, 95 hours, four days.

7

		Imperial	Metric
a	the speed of a car on the outside lane of a motorway in the UK	70–90 mph	110–145 kmph
b	the speed of someone having a brisk walk	3–5 mph	5–8 kmph
c	the speed of a top 100 m runner	20–25 mph	30–40 kmph
d	the speed of a supersonic jet aircraft	above 760 mph	above 1200 kmph

8 It is not possible. On the basis of ten people per m², the Isle of Wight will hold roughly 3.8 billion people.

9 a The rounded incomes are as follows.

Name	Income (rounded to the nearest £million)
National Trust	79
Oxfam	59
RNLI	56
Save the Children Fund	54
Imperial Cancer Research Fund	48
Cancer Research Campaign	45
Barnardos	36
Help the Aged	33
Salvation Army	32
NSPCC	31

b A horizontal barchart showing the earnings of the top five charities in the year in question.

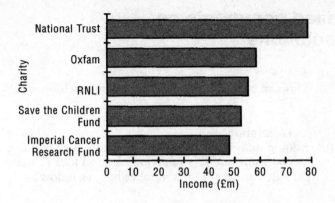

c A piechart would not be appropriate because the combined income from these five charities together, which would correspond to the complete pie, does not represent anything meaningful.

10 a Roughly £4.5 billion (I will accept anything between £4b and £5b.)
   b North America
   c Roughly £5 billion. This calculation assumes that the extra 10 million visitors spend at the same rate as do current visitors.

11 a Hamleys is the most expensive.
   b John Lewis and Toys Я Us are very similar in price, with Toys Я Us having the slight edge (a total price of £37.25, compared with John Lewis' £37.95).
   c Most expensive prices = £12.99 + £29.00 + £7.99 = £49.98
   Least expensive prices = £9.25 + £22.75 + £4.97 = £36.97
   Saving = £49.98 – £36.97 = £13.01
   d Percentage saving = $\frac{£13.01}{£36.97} \times 100$ = 35.2%

12 a The annual tax bill, T, can be found as follows.
   Subtract the tax allowances, A, from annual income, I, and multiply the result by 0.2.
   b (i) £490
     (ii) £9.42
   c £109.81.

Detailed comments on the solutions

1 a (i) 3×14;

This can be written out and calculated as follows:

$$\begin{array}{r} 14 \\ \times\ \underline{\ 3} \\ 42 \end{array}$$ Solution   42

(ii) Adding and subtracting with negative numbers can be confusing and it is sometimes a good idea to write the calculation out on a number line, as follows:

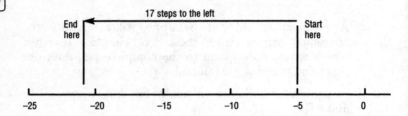

You start at -5 and then subtract 17. In other words, you move 17 steps to the left. This gives the result of -22.

(iii) $-20 \div 4$

Here you are dividing a negative number (-20) by a positive one (4). First you must decide on the sign of the answer (i.e. whether the answer is positive or negative). Because the two numbers are of different sign, the result must be negative, so write down the negative sign, $-$.

Next you do the calculation with the numbers.

$20 \div 4$ means $\frac{20}{4}$, giving the result 5.

So the solution is -5.

(iv) $15 \div 75$

This can be written as $\frac{15}{75}$.

The fraction can now be simplifed to the simplest equivalent fraction by dividing the numerator and the denominator by 15, giving the solution $\frac{1}{5}$ or 0.2.

A common mistake here is to do the division the wrong way round, giving $\frac{75}{15} = 5$.

b (i) $2\frac{1}{2} + 1\frac{1}{4}$

First add the whole numbers: $2 + 1 = 3$.

Next, add the fraction parts: $\frac{1}{2} + \frac{1}{4}$. Remember that in order to add fractions with different denominators, you must rewrite them as equivalent fractions which have the same denominator, which in this case is easiest done using quarters.

So, the fractions become $\frac{2}{4} + \frac{1}{4} = \frac{3}{4}$

Finally, add the fraction total to the whole number total;

Solution = $3 + \frac{3}{4} = 3\frac{3}{4}$.

(ii) $3\frac{1}{2} - 2\frac{3}{4}$

First subtract the whole numbers: $3 - 2 = 1$.

Write down what still has to be calculated: $1\frac{1}{2} - \frac{3}{4}$

Notice that you can't just subtract the fraction parts directly because the fraction being subtracted (three quarters) is bigger than the fraction it is being subtracted from (a half). The way round this is to borrow the whole number part, the 1, and turn it into quarters along with the fraction parts, as follows:

Solution = $1\frac{1}{2} - \frac{3}{4} = \frac{6}{4} - \frac{3}{4} = \frac{3}{4}$

(iii) $5\frac{1}{2} \times 3$

Multiply each part separately by 3 and then add the results together.

$5 \times 3 = 15$

$\frac{1}{2} \times 3 = 1\frac{1}{2}$

$15 + 1\frac{1}{2} = 16\frac{1}{2}$

(iv) $3\frac{2}{3} \times 5$

Multiply each part separately by 5 and then add the results together.

$3 \times 5 = 15$

$\frac{2}{3} \times 5 = \frac{10}{3} = 3\frac{1}{3}$

$15 + 3\frac{1}{3} = 18\frac{1}{3}$

c (i) 8^2, or 8 squared means $8 \times 8 = 64$

(ii) $\sqrt{81}$, or the square root of 81, means finding the number which, when squared, gives 81. The most obvious answer is 9, because it satisfies this condition – i.e. $9^2 = 81$. However, if you give the question a little further thought you may notice that there is another possible answer, namely, −9. You can check this by squaring −9, thus:

$(-9)^2 = -9 \times -9 = 81$

(iii) $\sqrt{(3^2 + 4^2)}$

First, work out what is inside the brackets:

$(3^2 + 4^2) = 9 + 16 = 25$.

Next, find the square root of 25. Using the same reasoning as in part (ii), this gives the two possible solutions, 5 or −5.

2 **a** $20 + 7 + \frac{3}{10} + \frac{6}{100} + \frac{4}{1000}$

These numbers have been arranged in a familiar pattern – tens, units, tenths, hundredths, and so on. Thus the number can be written down directly as 27.364.

b $60 - 3 + \frac{8}{10} + \frac{2}{100} - \frac{7}{1000}$

This question is similar to part **a** but slightly complicated by the two values which are subtracted. There is no single correct way of doing this, but my approach was to break it down as follows:

$60 - 3 = 57$

$\frac{2}{100} - \frac{7}{1000} = \frac{20}{1000} - \frac{7}{1000} = \frac{13}{1000} = \frac{1}{100} + \frac{3}{1000}$

Solution $= 57 + \frac{8}{10} + \frac{1}{100} + \frac{3}{1000} = 57.813$

c $\frac{62341}{1000}$

Division by 1000 has the effect of moving the decimal place three places to the left. The number 62341 has an invisible decimal point after the 1 (i.e. '62341.')

Thus, $\frac{62341}{1000} = 62.341$

3 The only comment here is that you need to keep in mind the sequence of decimal places, which are as follows:

… thousands hundreds tens units • tenths hundredths thousandths …

4 It is difficult to compare numbers written as fractions and the best strategy is to convert the fractions to either decimals or percentages.

a Converting three quarters to a percentage:

Three quarters as a percentage $= \frac{3}{4} \times 100 = 75\%$, which is larger than 70%. Incidentally, if you were unable to find three quarters of 100 in your head, use your calculator, as follows:

Press 3 ÷ 4 × 100 =

b Converting one twentieth to a decimal:

One twentieth as a decimal, $\frac{1}{20} = 0.05$, which is smaller than 0.06.

Alternatively, press 1 ÷ 20 =

c Converting two fifths to a decimal:

Two fifths as a decimal, $\frac{2}{5} = 0.4$, which is smaller than 0.5.

Alternatively, press 2 ÷ 5 =

d Converting an eighth to a percentage:

An eighth as a percentage $= \frac{1}{8} \times 100 = 12\frac{1}{2}\%$, which is larger than 10%.

Alternatively, press 1 $\boxed{\div}$ 8 $\boxed{\times}$ 100 $\boxed{=}$

e Converting a tenth to a percentage:
   A tenth as a percentage = $\frac{1}{10} \times 100$ = 10%, which is larger than 8%.
   Alternatively, press 1 $\boxed{\div}$ 10 $\boxed{\times}$ 100 $\boxed{=}$

5 There are no comments on this question except to suggest that you could develop your estimation skills by guessing some measures around the house, then getting out a tape measure, weighing scales, thermometer, and so on and checking how good your guesses were. It is surprising how quickly these skills do improve with practice.

6 As is the case for all questions about comparison of measures, the key thing is to convert all the measurements to the same units. Once this has been done, placing them in order of size becomes a trivial task. Suitable conversions are set out below for the measures, written here in the order in which they were originally given in the question.
   a 450 ml; 500 ml; 568 ml; 750 ml
   b 175 cm; 18 cm; 1 200 000 cm; 30 cm
   c 84 hours; 96 hours; 95 hours; 87.6 hours.

7 As with most estimation questions, there is no single correct method, as each person draws on their own past knowledge and experience.
   a My experience of motorway driving is that traffic on the outside lane seems to travel at around 80 mph (most drivers in the outside lane tend to break the speed limit of 70 mph unless there happens to be a police vehicle in the vicinity). So my first estimate here will be in these imperial units of miles per hour and then I will use my calculator (pressing 70 $\boxed{\times}$ 8 $\boxed{\div}$ 5 $\boxed{=}$) to convert to the metric equivalent, thus:
   70 mph = 70 $\times \frac{8}{5}$ kmph = 112 kmph. I rounded this to 110 kmph.
   90 mph = 90 $\times \frac{8}{5}$ kmph = 144 kmph. I rounded this to 145 kmph.
   b As with the previous part, I know from experience that 4 mph represents a fairly brisk walk, so I allowed a range of between 3 and 5 mph. The conversions to kmph were done as above in part **a**.
   c This time I had no idea how fast a top sprinter could run, so I decided to do a calculation instead. Again, drawing on my past experience, I know that a good time for the 100 m is around 10 seconds. These seem to be convenient

numbers, so I chose to work in metric units this time and will convert to imperial afterwards.

In 10 seconds, the sprinter travels 100 m

In 1 minute, the sprinter travels 100 × 6 m

In 1 hour, the sprinter travels 100 × 6 × 60 m = $\frac{100 \times 6 \times 60}{1000}$ km.

Pressing the calculator sequence 100 ⊠ 6 ⊠ 60 ⊡ 1000 ⊡ gives the answer 36. In other words, the sprinter's speed is 36 kmph.

An aside on cancelling out fractions

There is an alternative method of working out this last answer which involves 'cancelling' out the fraction. In general, cancelling out a fraction means dividing numbers in the top and bottom parts of the fraction by factors that they have in common. This has the effect of simplifying the fraction before it is evaluated. In the case of the fraction $\frac{100 \times 6 \times 60}{1000}$, it means dividing out the tens and hundreds. This has been shown in separate stages below.

First, divide top and bottom of the fraction by 100, giving the following:	$\dfrac{\overset{1}{\cancel{100}} \times 6 \times 60}{\underset{10}{\cancel{1000}}}$
There is still further scope for cancelling, so divide the 60 on the top and the remaining 10 on the bottom by 10, thus:	$\dfrac{\overset{1}{\cancel{100}} \times 6 \times \overset{6}{\cancel{60}}}{\underset{1}{\cancel{10}}}$
This can be tidied up and simplified as follows:	$\dfrac{1 \times 6 \times 6}{1} = 36$

So, the result, as before, is 36 kmph. Again, if you wanted to use your calculator for this calculation, press 100 ⊠ 6 ⊠ 60 ⊡ 1000 ⊡.

On the basis of this figure of 36 kmph, I allowed you to mark yourself correct if your answer fell within the range 30–40 kmph.

(Incidentally, in practice no sprinter could possibly sprint at this speed for an hour, but calculating someone's speed in mph or kmph does not necessarily imply that they continued to travel for an hour, even though my wording above suggests that they do.)

Finally, I can convert to mph as follows.

30 kmph = 30 × $\frac{5}{8}$ mph = 18.75 mph. I rounded this to 20 mph.

Alternatively, press 30 $\boxed{\times}$ 5 $\boxed{\div}$ 8 $\boxed{=}$

40 kmph = 40 × $\frac{5}{8}$ mph = 25 mph,

which required no further rounding.

Alternatively, press 40 $\boxed{\times}$ 5 $\boxed{\div}$ 8 $\boxed{=}$

d This again was a fact that I happened to have stored away in my brain. And, having been brought up from childhood with imperial units, I remembered that the speed of sound is around 760 mph. If you had absolutely no idea, try looking up the word 'sound' in a dictionary or encyclopedia. There is no need for any greater accuracy than this because the speed of sound varies, depending on such things as the nature of the gas that it is passing through, the air temperature at the time, and so on. As before, the convertion to kmph is easy with a calculator.

Press 760 $\boxed{\times}$ 8 $\boxed{\div}$ 5 $\boxed{=}$

I rounded the calculator result of 1216 kmph to 1200 kmph.

Supersonic aircraft travel at speeds greater than 760 mph or 1200 kmph. But clearly there needs to be a sensible upper limit to your answer – say, 2000 mph or 3000 kmph.

8 Number of people who fit into 1 m^2 = 10
Number of people who fit into 1 km^2 = 10 × 1 000 000
(Remember that there are 1000 m in one km but 1000 × 1000 = 1 000 000 m^2 in 1 km^2)
Number of people who fit into
381 km^2 = 10 × 1 000 000 × 381
        = 3810 million, or
        3.81 billion

Incidentally, the claim that 'it is possible to fit the world's population on the Isle of Wight' may not be plausible today, but was probably more valid when it was first thought up many decades ago. At present rates, the world's population is doubling roughly every 60 years. This calculation is explained more fully in Part Two of this book (page 214).

9 There are no additional comments on this question.

10 a The task here is to try to estimate what fraction each slice is of the total. You can see that the slice corresponding to Europe takes up almost half of the pie, so the annual spending represented by this slice would be almost half of £9.2b, or roughly £4.5b.

b A spending of £2b out of a total spending of £9.2b represents the following fraction of the pie:

$$\frac{2}{9.2}$$

On my calculator, this gives a decimal value of just over 0.2, or roughly one fifth. So I am looking for a slice which is slightly bigger than one fifth of the pie. Only North America fits the bill here. (If you imagine four more slices the same size as North America, it seems reasonable that five of these slices would together make a complete pie.)

c This final part is an exercise in proportion. We know that:

19.2 million visitors spent £9.2b.

So, 1 million visitors should spend $\frac{£9.2b}{19.2}$

Then 10 million visitors should spend $\frac{£9.2b}{19.2} \times 10$.

On the calculator, press 9.2 $\boxed{÷}$ 19.2 $\boxed{×}$ 10 $\boxed{=}$ giving the result 4.7916666, which I rounded to £5b.

11 a Since the prices in Hamleys were the highest for each of the three toys listed, this first question was easy to answer.

b There isn't an obvious method for answering this question, but I decided to calculate the total price of all three items and select as the cheapest the store with the smallest total, which was Toys Я Us. However, the price differences between John Lewis and Toys Я Us are so small that John Lewis could possibly come out cheapest if three different toys were chosen. Based only on data from three toys, it is impossible to come up with a clear answer to this question.

c and d There are no additional comments on these parts.

12 a There are no additional comments on this part.

b (i) First you must subtract Petra's tax allowances from her annual income. On the calculator, this is done as follows:

6200 $\boxed{-}$ 3750 $\boxed{=}$

giving the result 2450.

Next multiply the result by 0.2. There is no need to re-enter the 2450 as it is already on the calculator display, so simply press $\boxed{\times}$ 0.2 $\boxed{=}$

(ii) The previous result of £490 should still be on your calculator display. This is the annual tax bill. To calculate this in weekly terms you must divide by 52, so simply press $\boxed{\div}$ 52 $\boxed{=}$, giving the result 9.4230769. I rounded this to the nearest penny, giving £9.42.

c Petra's weekly earnings net of tax can be calculated as follows.

Her annual earnings net of tax: Press 6200 $\boxed{-}$ 490 $\boxed{=}$
Her weekly earnings net of tax: Press $\boxed{\div}$ 52 $\boxed{=}$
giving the result 109.80769, which I rounded to the nearest penny, giving £109.81.

part two

appendices –

mathematics in action

Appendix A: Calculating a best buy

Whether you are buying potato crisps, rice or hair shampoo, most supermarket purchases are made available in packs of different size and price. Sometimes you choose the size for practical reasons – here are some examples.

- 'The larger toothpaste tubes always fall out of our bathroom mug so I tend to buy the smallest one.'
- 'In our household a small pack of Cornflakes lasts about two days so I always buy the largest one.'

But in many situations you simply want to buy the size which gives the best value for money.

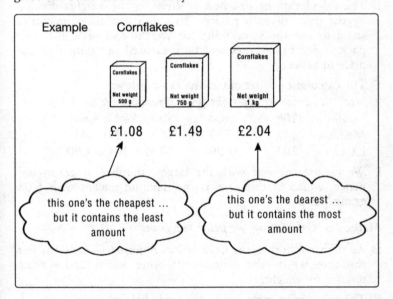

Note: You can't compare the prices directly because each packet contains different amounts of Cornflakes. You need to find a way of comparing like with like. There are two possible methods.

For each packet:

a calculate the weight of Cornflakes per pence and then choose the packet which works out at the largest weight per pence.
b calculate the cost in pence of Cornflakes per gram and then choose the packet which works out cheapest per gram.

Note: In method **a** you end up choosing the packet which produces the largest answer to your calculation, while method **b** involves choosing the packet which produces the smallest answer. Remember that your calculator will do the arithmetic. All you have to do is three things, summarized by the letters DPI; in other words:

- **D** *decide* what calculations to do and understand why you are doing them
- **P** *press* the right buttons in the correct order
- **I** *interpret* the answers sensibly.

Method **a** *Calculating the weight per pence*

D The calculation needed here is division; the weights divided by the price of each packet. To avoid confusion, it makes sense to use the same units for weight and price for each packet. So the weights will be measured in grams and the price in pence.

P The calculation is set out in the table below.

Size	Price (p)	Weight (g)	Grams per p (to 3 figures)
Small	108	500	$500 \div 108 = 4.63$
Medium	149	750	$750 \div 149 = 5.03$
Large	204	1000	$1000 \div 204 = 4.90$

I We choose the size with the largest number of grams per pence, which in this case is the medium packet with 5.03 grams per p.

Method **b** *Calculating the price per gram*

D As with method **a**, the calculation required is division, but this time we do the division the other way round – price divided by weight.

P The calculation is set out in the table below.

Size	Price (p)	Weight (g)	Pence per g (to 3 figures)
Small	108	500	$108 \div 500 = 0.216$
Medium	149	750	$149 \div 750 = 0.199$
Large	204	1000	$204 \div 1000 = 0.204$

I This time we are looking for the size with the cheapest price per g. As before, we select the medium packet with 0.199 pence per g.

Points to note in value-for-money calculations
- It doesn't matter which method you use (calculating weight per pence or price per gram). You just need to be clear which one you have chosen and make sure to use the same method throughout.
- Ensure that the units of measure match up – don't calculate one pack size priced in pence and another in pounds.
- Whichever method you adopt determines whether you will choose the packet whose calculation yields the *largest* value or the *smallest* value. Remember that you want to pay small pence and you want to receive large quantities. Thus

For the calculation … you want … so you choose

| grams per p | large grams | the largest one |
| pence per g | small pence | the smallest one |

- We have only looked at a simple example where the goods being compared were identical in every respect except size and price. For most purchases, there are many other factors to take into account when deciding on value for money and it is altogether more complicated than these calculations suggest. For example you may also wish to take account of quality, durability, prestige, recyclability, and so on; all factors that are much more difficult to measure and calculate with.
- Unit pricing is a practice followed by most supermarkets. As well as including on the label of each item the price and the size (weight or capacity, as appropriate), the 'unit price' is also included. This allows you to compare the relative value of products across different pack sizes. Here are some examples.

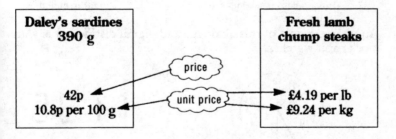

Notice that when supermarkets unit price, they usually reduce the price to a *suitable* unit, not necessarily to a single gram or pound. In the examples above, the basic unit for sardines was taken to be 100 g, while that for lamb steaks is kg. The reason for this is to avoid having to use prices written as awkward decimal numbers that people find hard to make sense of. Note also, meat has been unit priced in both imperial and metric units for the customer's convenience.

Appendix B: Reading the 24-hour clock

Analogue and digital

The world is increasingly divided into two types of device – analogue and digital. Analogue devices are so called because of their mechanical way of working: the mechanism is the device. For example, a vinyl record player produces musical sound in a mechanical way in that, as the record spins, the needle moves about inside the grooves. That movement is then translated into sound. Contrast that with digital sound where a laser reading device merely scans a long list of numbers (i.e. digits) encoded in the compact disc or digital audio tape. It is these numbers that are then transformed into sound.

a record is analogue   a CD is digital

Another example of this analogue and digital distinction is with clocks and watches.

an analogue clock   a digital clock

The old-fashioned analogue clocks tend to have a circular face numbered 1 to 12 and hands that sweep round, marking out the time. As with the record player, there is a moving mechanism that physically marks out a circular path which we interpret in terms of time passing. One complete circuit of the clock face by the hour hand represents the passing of 12 hours. Two circuits of the clock face by the hour hand gives a full day of 24 hours. We use common sense to distinguish between morning time (a.m.) and afternoon time (p.m.).

Digital clocks and watches, on the other hand, simply produce numbers. These numbers can be organized in twelve hour cycles, in which case the letters a.m. or p.m. are shown on the display. Alternatively, most digital clocks can be set to display time in cycles of 24 hours. The chart below shows how 12 hour and 24 hour times are related.

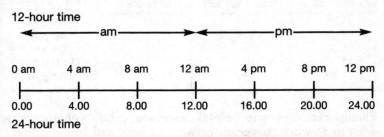

As you can see, for the first 12 hours in a day (i.e. during the a.m. period) the 12-hour and 24-hour times are exactly the same. However, after 12 a.m., the times roll back to zero on the 12-hour system, whereas they simply continue (13, 14, 15, ...) on the 24-hour system.

Converting from 12-hour time to 24-hour time

Remember that 24-hour time is a measure of how long it is since the previous midnight. So...
...if it is a.m., the 24-hour and the 12-hour times are the same, and
...if it is p.m., you have to add 12 hours.
Here are some examples.

12-hour time	a.m. or p.m.?	+ 12 hours?	24-hour time
3.15 a.m.	a.m.		3.15
6.44 p.m.	p.m.	+ 12.00	18.44
4.52 p.m.	p.m.	+ 12.00	16.52
5.00 a.m.	a.m.		5.00
11.07 p.m.	p.m.	+ 12.00	23.07

Converting from 24-hour time to 12-hour time

Remember that any time after 12 noon is p.m., and for afternoon times the 12-hour clock rolls back to zero. This means that, if the 24-hour time is greater than 12 (i.e. if it is a p.m. time), you must subtract 12 to find the 12-hour time.

Here are some examples.

24-hour time	more than 12?	– 12 hours?	12-hour time
13.55	yes	– 12.00	1.55 p.m.
16.40	yes	– 12.00	4.40 p.m.
4.08	no		4.08 a.m.
11.50	no		11.50 a.m.
21.33	yes	– 12.00	9.33 p.m.

Finally, here are some 'helpful' diagrams to help you sort out what to do when converting between 12-hour and 24-hour time.

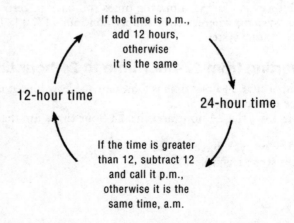

If the time is p.m., add 12 hours, otherwise it is the same

12-hour time

24-hour time

If the time is greater than 12, subtract 12 and call it p.m., otherwise it is the same time, a.m.

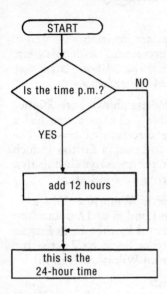 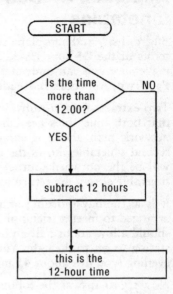

Why?

Can't we stick to the good old-fashioned 12-hour clocks?

The chief virtue of the 24-hour system is that it automatically does away with the need to specify whether the time is a.m. or p.m. For most everyday purposes, this may not seem very important, but when consulting train, bus and airline timetables, it makes sense to use a system which is not prone to confusion. It has been estimated that, over the first ten years of the introduction of the 24-hour clock in their timetables, British Rail staff costs fell by nearly £8 million in today's terms, due to dispensing with the need to pursue interminable conversations with customers along the lines of, 'Excuse me. Is that 4 o'clock a.m. or 4 o'clock p.m.?' etc.

(Actually, I just made that last 'fact' up, but you get the general point!)

Appendix C: Bus and railway timetables

Bus, railway and aeroplane timetables are invariably written in terms of the 24-hour clock. Before proceeding with this item, make sure you can understand and use the 24-hour clock (see Part Two, Appendix B: Reading the 24-hour clock).

Two extracts from a railway timetable are shown here. Notice that both timetables are labelled Table 5 but each one has a network map above it showing the direction of travel. The second timetable shows the journey *to* London Euston (which will be the outward journey for our purposes) while the first timetable covers the return journey *from* Euston station.

It is a Tuesday morning and you are in Wilmslow. You have arranged to meet a friend in a cafe in London at 12 noon. You should allow about half an hour to travel by tube from Euston station to get to the cafe. You want to be home by 7 p.m. that evening (you live about 45 minutes from Wilmslow station).

Now try to answer the following questions.
a What train should you catch if you want to be certain of arriving at the cafe before your friend? How long are you likely to have to wait at the cafe if you catch this train?
b What is the most sensible train to catch?
  • Is this a direct service or will you have to change trains?
  • If you have to change, where are you likely to have to change and how long may you have to wait for that connection?
  • At what time would you estimate arriving at the cafe?
c What train will you catch to return home? Will you be able to eat on this train?
d What are the train journey times each way?

Table 5 INTERCITY West Coast

Full service London — Rugby Table 4.   Full service London — Crewe, Table 6.

	🍽A	🍽S	✗	✗	✗	✗	✗	✗	🍽	✗
Mondays to Fridays										
London Euston	0655	0800	0900	0910	1000	1020	1100	1200	1300	1310
Watford Junction	0711u	0816u	—	0926u	1016u	1037u	—	1216u	1241	1326u
Milton Keynes Central	0734	—	0936	—	—	—	1135	—	1334	—
Rugby	—	—	—	1011	—	—	—	1301	—	1411
Stoke-on-Trent	0844	—	1046	1130g	1144	—	1244	—	1444	1531g
Macclesfield	0903	—	1105	1152g	1203	—	1303	—	1503	1549g
Crewe	—	—	—	1112	—	1222	—	—	—	1512
Wilmslow	—	1010	—	1204	—	1304	—	1414	—	1604
Stockport	0918	1020	1120	1207g	1218	—	1318	1424	1518	1605g
Manchester Piccadilly ↓	0932	1033	1132	1219g	1234	—	1332	1437	1532	1619g

	🍽	✗	✗	🍽S	✗	🍽S	✗	🍽S	✗	✗
				A				A		
Mondays to Fridays										
London Euston	1400	1410	1500	1600	1630	1700	1735	1800	1805	1900
Watford Junction	1416u	—	—	1616u	1646u	—	—	1816u	1822u	—
Milton Keynes Central	—	1444	1534	—	—	—	—	—	1845u	—
Rugby	—	1511	—	—	—	—	1834	—	1912	—
Stoke-on-Trent	1544	1619g	1644	—	1814	—	1938	1950	—	2045
Macclesfield	1603	—	1703	—	1833	—	1957	2009	—	2106
Crewe	—	1612	—	—	—	—	—	—	2008	—
Wilmslow	—	1704	—	1808	—	1913	—	—	2104	—
Stockport	1618	1716	1718	1818	1848	1923	2012	2024	2116	2123
Manchester Piccadilly ↓	1636	1728	1732	1834	1902	1936	2025	2037	2128	2137

	🍽	🍽			🍽	🍽	🍽	🍽	🍽	🍽
Mondays to Fridays	✗			**Saturdays**						
London Euston	2000	2200	—	—	0645	0730	0750	0800	0850	0900
Watford Junction	—	2216u	—	—	0701u	—	0808u	—	0824	0916u
Milton Keynes Central	2034	2239	—	—	0724	0805	—	0834	0924	—
Rugby	—	—	—	—	—	0844	—	—	—	1011
Stoke-on-Trent	2144	—	—	—	0844	—	0944	—	1046	1130g
Macclesfield	2203	—	—	—	0903	—	1003	—	1105	1152g
Crewe	—	0020	—	—	—	0942	—	1012	—	1112
Wilmslow	—	0042g	—	—	—	1016	—	1104	—	1204
Stockport	2218	0102g	—	—	0918	—	1018	—	1120	1207g
Manchester Piccadilly ↓	2232	0114	—	—	0932	—	1034	—	1133	1219g

	🍽	🍽	🍽
Saturdays			
London Euston	0950	1000	1050
Watford Junction	1008u	—	—
Milton Keynes Central	—	1034	1125
Rugby	—	1111	—
Stoke-on-Trent	1144	—	1244
Macclesfield	1203	—	1303
Crewe	—	1212	—
Wilmslow	—	1304	—
Stockport	1218	1316	1318
Manchester Piccadilly ↓	1234	1329	1332

On-Board Services (see page 1):
🍽 InterCity train with catering.
🍽 First Class Pullman.
✗ Restaurant.
🍽 Restaurant (First Class only).
S Silver Standard.
🍽 Seat reservations essential free of charge.

Notes for this and opposite page:
A   The Manchester Pullman.
C   Until January 8.
D   Until Jan 8; then from Apr 16.
E   Until Jan 7/8; also May 27/28.
G   Jan 14/15 to May 20/21.
H   From January 15.
J   January 15 to April 9.
g   Change Stafford.
k   0044 May 28 only
    (Sunday morning).
n   0120 May 28 only
    (Sunday morning).
p   Until Jan 8; then from Apr 16
    at this station.
q   Jan 15 to Apr 9 at this station
    by special bus from Crewe.
s   Calls to set down only.
u   Calls to pick up only.

Source: A Guide to Intercity Services © British Railways Board

Light printed timings indicate connecting service.

Table 5 INTERCITY West Coast

Stockport · Macclesfield · Stoke-on-Trent
Manchester Piccadilly · · · · · London Euston
via Crewe · Watford

Full service Crewe — London, Table 6. Full service Rugby — London, Table 4.

Mondays to Fridays

			A	A						
Manchester Piccadilly	0520	0600	0645	0710	0730	0630	0833	0930	1030	1033
Stockport	0528	0608	0654	0718	0738	0638	0842	0938	1038	1042
Wilmslow	0536	—	—	0727	—	—	0850	—	1046	1050
Crewe	0557	—	—	—	—	—	0925	—	—	1125
Macclesfield	—	0622	0706	—	0752	0852	—	0952	—	1039g
Stoke-on-Trent	—	0842	0727	—	0812	0912	—	1012	—	1102g
Rugby	—	0730	—	—	—	1023	—	—	—	1223
Milton Keynes Central	—	0824	—	—	0924	—	1049	—	1221	1247
Watford Junction	—	0856	—	0921s	1013	1045s	1129	1142s	1313	1329
London Euston	0813	0840	0916	0944	1011	1108	1136	1205	1307	1334

Mondays to Fridays

									A	
Manchester Piccadilly	1130	1133	1230	—	1330	1333	1430	1433	1530	—
Stockport	1138	1142	1238	—	1338	1342	1438	1442	1538	—
Wilmslow	—	1150	—	1250	—	1350	—	1450	—	1527
Crewe	—	1225	—	1325	—	1425	—	1525	—	1608
Macclesfield	1152	—	1252	—	1352	—	1452	—	1552	—
Stoke-on-Trent	1212	—	1312	—	1412	—	1512	—	1612	—
Rugby	—	1325	1359	—	—	1527	—	1628	—	1725s
Milton Keynes Central	—	—	1424	1447	—	—	1618	1653	—	1725s
Watford Junction	1342s	1412s	1513	1529	1542s	1614s	1659	—	1742s	1753s
London Euston	1405	1435	1510	1534	1605	1637	1705	1739	1805	1816

Mondays to Fridays

	A							
Manchester Piccadilly	1630	—	1730	—	1830	—	1933	2000
Stockport	1638	—	1738	—	1838	—	1942	2008
Wilmslow	1648	—	—	1734	—	1825	1950	—
Crewe	—	1723	—	1823	—	1915	2026	—
Macclesfield	—	1640g	1752	—	1852	—	—	2022
Stoke-on-Trent	—	1700g	1812	—	1912	—	—	2042
Rugby	—	1823	—	1921	—	2134	—	—
Milton Keynes Central	1820	1847	1918	—	2018	—	2159	2204
Watford Junction	1856	—	1947s	2010s	2113	—	2227s	2232s
London Euston	1905	1934	2010	2033	2105	2114	2250	2305

Saturdays

Manchester Piccadilly	0530	0630	0700	0730	0630	0833	0930	1030	1033	1130
Stockport	0538	0638	0708	0738	0638	0842	0938	1038	1042	1138
Wilmslow	—	—	0716	—	—	0850	—	1046	1050	—
Crewe	—	—	0736	—	—	0925	—	—	1125	—
Macclesfield	0552	0652	—	0752	0652	—	0952	—	1039h	1152
Stoke-on-Trent	0612	0712	—	0812	0912	—	1012	—	1102h	1212
Rugby	—	—	—	—	—	1023	—	—	1223	—
Milton Keynes Central	—	—	0820	—	0924	—	1049	—	1221	1247
Watford Junction	—	0755s	0856	0823s	—	1045s	1129	1142s	1313	1329	1342s
London Euston	0828	0917	0856	1021	1118	1146	1215	1317	1344	1415

Notes for this and opposite page:

A The Manchester Pullman.
g Change Stafford.
h Change Stafford.
  (May 27 only Macclesfield 1037 Stoke-on-Trent 1056.)
p Until January 8 at this station by special bus to Crewe.
q 1752 Jan 15 to Apr 9 by special bus to Crewe.
s Calls to set down only.
u Calls to pick up only.

On-Board Services (see page 1):

InterCity train with catering.
First Class Pullman.
Restaurant.
Restaurant (First Class only).
Silver Standard.
Seat reservations essential free of charge.

Light printed timings indicate connecting service.

Source: A Guide to Intercity Services © British Railways Board

Solutions

a For the outward journey, you need the second timetable. The 0727, which gets into London Euston at 0944 should get you to the cafe by 1015 – an hour and three quarters before the agreed time, so not very satisfactory!

b It's a bit tight, but you should just make your assignation if you catch the next train from Wilmslow, the 0850, getting into London Euston at 1136.

 • The 0850 departure is not a direct service. You can tell this because of the light printing of the departure time of 0850. As you can see from the explanation at the bottom of the page, 'Light printed timings indicate connecting service', so this will require a change of trains.

 • In this case you will have to change at Crewe, which is the next main station. Note that the times 0850 (in light printing) and 0925 (in bold) are the *departure* times from Wilmslow and Crewe, respectively. Since you will be changing at Crewe, you will expect to arrive there some time before the 0925 departs. You can make an intelligent guess at your arrival time in Crewe by looking at a previous column of figures. Notice that the 0536 from Wilmslow gets into Crewe at 0557, suggesting a journey time of 21 minutes. Assuming the 0850 travels at the same speed, it should get into Crewe by 0911, allowing you ample time (14 minutes, in fact) to make your connection on the 0925.

 • This should get you to the cafe just a few minutes after 12 noon.

c To choose the homeward train, you need to work backwards from when you want to get home, as follows:
Getting home by 7 p.m. means arriving at Wilmslow station by 6.15 p.m., i.e. by 1815. According to the first timetable, there is an ideal train which departs from London Euston at 1600 and gets into Wilmslow at 1808. *Note*: According to the codes at the top of the column, this train is a First Class Pullman (called the Manchester Pullman – see note A on the timetable) with a Silver Standard, but with no Restaurant facilities. So, you will be able to eat on this train, but not in high style!

d The journey times are shown in the table below

Depart		Arrive		Journey time
Wilmslow	0850	London Euston	1136	2 hours 46 mins
London Euston	1600	Wilmslow	1808	2 hours 08 mins.

So, the return journey is quicker by some 38 minutes.

You may have gone wrong calculating journey times on your calculator. For example, the outward journey ran from 0850 to 1136, so pressing 1136 ⊟ 0850 ⊟ gives 286 (i.e. 2 hours and 86 mins) and not 246 (or 2 hours 46 mins) as shown above.

Calculating journey time cannot easily be done on a calculator. The complication is that there are 60, not 100 minutes in one hour. There are many common sense ways of tackling this problem. Here is how I worked out the journey time between 0850 and 1136.

- First, I added ten minutes to 0850 to bring it up to the next exact hour (0900). This shortens the journey time by ten minutes, but I'll add it on later.
- Next, I calculated the journey time from 0900 to 1136 – this is easy to do in your head, the answer being 2 hours 36 mins.
- Finally, I need to remember that I shortened the journey time by ten minutes, so I must now add these on.
  So, 2 hours 36 mins + 10 mins gives the answer, 2 hours 46 mins.

Appendix D: Checking the supermarket bill

Most people simply haven't got the time or the energy to check their weekly supermarket bill item by item. In general, we tend to assume that the machine has got it right. Where errors occur, sometimes they are to the customer's advantage and sometimes to the store's advantage, but it is likely that, taken over the long term, errors tend to average themselves out.

Typically, the item will be scanned at the checkout with a barcode reader. Occasionally the barcode is read incorrectly but barcodes have a built-in 'checksum' that bleeps when there is an error – this is explained on pages 202–205. However, errors do occur, and it is worth being awake to that possibility at the checkout to avoid it costing you money. The key thing is to know when to check the bill in detail and how to do so if you had to. Here are a few guidelines.

What does my bill usually come to?

First of all, it is helpful to know roughly what your bill is likely to come to. If you do a regular weekly shopping at the same store, you may be able to do this with some accuracy. For example, your typical bill may come to, say, around £55, so anything less than £40 or more than £70 in a particular week ought to make you suspicious.

Are there any unusual and expensive items this week?

If you have bought some untypical and expensive items (for example, alcohol or kitchen or electrical goods), make an estimate of these and add them to your typical bill. That way you will know what you can expect your bill to come to, roughly.

How many items did I buy?

Most supermarket receipts state the total number of goods bought. In a recent large shopping spree for groceries, I spent about £90, having bought 73 items. This works out at just over £1 per item. For most supermarket bills, an average of about £1 per item is fairly typical and using this fact might provide you with a quick check that the overall bill is in line with the contents of your trolley. Thus, buying, say, 40 items, I might expect my bill to be around £50–£60. This strategy might allow you to pick up situations where £69 was recorded instead of 69p, say, for a bag of apples.

Can I do a quick mental check of the bill?

If there is a huge number of items on your bill – as many as 73, for example – it isn't realistic to do a mental check. However, if there are, say, only 11 or 12 items, then this is perfectly possible. There is no single correct method of rounding – you could round to the nearest 10 pence or 50 pence. The method below is more approximate and is based on rounding the prices to the nearest £.

P PERKINS PLC

THE BUTTS
WARWICK
CV21 3FL

Telephone no. 01926 334215

	£
Custard powder	0.85
Apple jce 2L	1.69
PP County spread	1.19
PP H-gran stick	0.56
Baking potatoes	1.98
PP Earl Grey tea	0.69
PP Earl Grey tea	0.69
PP Lentils 500g	0.59
Orange jce 1L	1.89
PP Eggs small	0.77
Baked beans	0.23
11 Bal Due	11.04

4356 601364 645355431

The method is based on looking at the pence part of the price. Where the amount of the pence is 50p or more, round up to the next whole number of pounds, othewise ignore the pence, which has the effect of rounding down the pounds. For example, the first item, which is 85p will be rounded up to £1 because the pence (85p) is greater than 50p. On the other hand, a sum of £1.19 is rounded down to £1 because the pence (19p) is less than 50p.

Adopting this procedure, the costs of the eleven items are rounded as follows:

Item	Actual price (£)		Rounded price (£)
Custard powder	0.85	→	1
Apple jce 2L	1.69	→	2
PP County spread	1.19	→	1
PP H-gran stick	0.56	→	1
Baking potatoes	1.98	→	2
PP Earl Grey tea	0.69	→	1
PP Earl Grey tea	0.69	→	1
PP Lentils 500g	0.59	→	1
Orange jce 1L	1.89	→	2
PP Eggs small	0.77	→	1
Baked beans	0.23	→	0
	Rounded Total		£13

In this particular case, the rounding method has produced an answer which is too high – by roughly £2. However, the object of the exercise has been to produce a rough 'order of magnitude' answer in order to check whether the total is more or less correct, rather than to get a precise answer.

Appendix E: Understanding a shop receipt

If you ever fancied yourself as a latter day Poirot or Sherlock Holmes, you could do a lot worse than to practise your skills uncovering the hidden mysteries of a lowly shop receipt. You may be surprised to discover how much you can tell about a person simply by rummaging around in their discarded plastic bags and fishing out the sordid details of their last shopping transaction – which might look something like this!

S DEVLIN PLC
THE SHIRES PARK
DODDINGTON
CV11 4RR

TELEPHONE NO. 01926 43526

SALES VOUCHER: CUSTOMER'S COPY

0873456	D/WASH LIQ LEM	3.29
0786543	TULIPS	2.95
2 BAL DUE		6.24
CASH		10.00
CHANGE		3.76

503 24 1032   16:26   24DEC01

VAT NO. 534 5644 89

**THANK YOU FOR SHOPPING WITH
DEVLIN**

Please retain your receipt

As you can see, this piece of evidence tells its own story about the transaction that took place. You might like to reconstruct part of that story, sorting out in your mind the objective facts involved – the 'where', 'when' and 'what' of the transaction. To help you, try completing the blanks in the police file on the next page.

Police file

We have reason to believe that the supect entered the premises

of _____ (shop) at _____ (address) in

the town of _____ on the afternoon of _____ in

the year _____. At precisely _____, ____items were

purchased, namely a _____ and a _____,

costing £_____and £_____respectively. A £_____ note

was submitted to the cashier and £_____ in change was

received.

For the completed file turn to page 216.

We might like to go a little further here and speculate what sort of person this was. Note the date, which was Christmas Eve. Now most people who celebrate this festival are still frantically buying the basics by Christmas Eve (in a few homes the turkey, crackers, balloons, presents ... have still to be bought). This individual has clearly got the whole Christmas thing totally under control if he or she is making a special trip for tulips and dishwasher liquid on Christmas Eve! Also, assuming that they own the dishwasher in question, you might suppose that they aren't exactly in the bottom income bracket.

In short, one has a picture of someone who is reasonably well off, certainly well organized, who wants to relax this Christmas with no plans to be hand-washing dishes in the sink!

Follow up

You might like to find various tickets and receipts of your own and see if you can crack all the codes they contain. I have focused on what information they provide for the customer. What additional information do they provide for the management? How might they be used for stock control?

Appendix F: Checking the VAT

Value added tax (VAT) is charged on many of the goods bought in the UK. At the time of writing, the rate of tax is 17.5%. What this means is that, for every £100 net value, the VAT charge is £17.5, bringing the total to £117.5. In other words:

Net value		VAT		Gross value
£100.00	+	£17.50	=	£117.50

This is what the customer pays

Below is a simplified receipt from a plumbing centre where I recently bought the wherewithal to install a ventilator into an external wall of my kitchen. (In the event I failed disastrously to complete the job without professional help, but that's another story!)

Catalogue No.	Qty. Ordered	Description	VAT	Price	Per	Total net
800160	1	Stadium BM720 black hole ventilator	17.5	28.13		28.13
602047	1	Supaset rapid set cement – 3kg	17.5	4.46		4.46
					Sub total	32.59
					VAT	5.70
					Total	38.29

Let's check that the basic arithmetic is correct.

a Is the subtotal of £32.59 correct?
From the right hand column, we can see that the two items cost £28.13 and £4.46, respectively. Pressing 28.13 ⊞ 4.46 ⊜ on the calculator confirms the answer given in the Sub total box, £32.59.

b Is the VAT of £5.70 correct?
Notice that in this receipt the net totals are added first and then the overall VAT is calculated at the end. In order to check the VAT, you must find 17.5% of 32.59. As was explained in Chapter 06, this is found by converting 17.5% to its decimal form (giving 0.175) and multiplying this by 32.59, thus:
0.175 ⊠ 32.59 ⊜
giving an answer 5.70325.
Rounding this answer to the nearest penny gives the result 5.70, i.e. £5.70, which confirms the value in the VAT box.

c Is the total of £38.29 correct?
Adding the net subtotal and the VAT should give the overall gross total, thus:
32.59 ⊞ 5.70 ⊜
confirming the final bill of £38.29.

Some additional questions

Surely it's not necessary to work out the VAT on its own if I simply want to check that the overall bill, inclusive of VAT, is correct?

You are quite correct – it is not necessary to find the VAT first and then add it on! The VAT-inclusive bill can be found directly by multiplying the net bill by 1.175, thus:

1.175 $\boxed{\times}$ 32.59 $\boxed{=}$

Again, after rounding, you should find that this calculation confirms the final bill of £38.29.

My calculator has a percentage key marked on one of the buttons. How can I use it to work out VAT?

Unfortunately not all calculator percentage keys are designed to work in the same way. Indeed, some seem to operate in a most bizarre way! You may need to consult your calculator manual to check this for yourself. However, here are some suggestions for things to try. Incidentally, if you are trying things out on a calculator to see how it works, choose simple numbers! I suggest that you try to add, say, 8% on to 100, knowing that the answer should be 108.

Try pressing these sequences and see what you get

     100 $\boxed{+}$ 8 $\boxed{\%}$
     100 $\boxed{+}$ 8 $\boxed{\%}$ $\boxed{=}$
     100 $\boxed{\times}$ 8 $\boxed{\%}$
     100 $\boxed{\times}$ 8 $\boxed{\%}$ $\boxed{=}$

On some calculators you will simply not get a satisfactory result to this calculation. For example, on one of my calculators, the $\boxed{\%}$ key has been set up solely to convert fractions to percentages, thus:

3 $\boxed{\div}$ 5 $\boxed{\%}$ produces the answer 60, because the fraction $\frac{3}{5}$ = 60%.

Overall, then, the percentage key is a bit of a mixed blessing! It may be useful in VAT calculations, but, provided you understand how to convert a percentage to a decimal, you really don't need such a key.

Appendix G: Cooking with figures

As a child living in Ireland, I remember watching my grandmother baking soda bread on a griddle. I asked her how she did it.
'Well,' she said, 'you start by taking two gopins of flour ...'
She then had to explain to me that a 'gopin' was a double handful.
'But how do you know when you've got exactly a gopin?,' I asked.
'Oh, you just know by the feel of your hand,' she replied.

Measurements in cooking these days tend to be much more precise. Recipes are usually stated in formal units like grams, lbs, litres, and so on and these were explained in Chapter 07, Measuring. This section covers one or two specific questions that often present themselves in the kitchen which require some mathematics.

How do teaspoons, pints and litres match up?

Most recipes published in the UK tend to be stated both in imperial units and metric units, as well as in more informal units such as teaspoons, tablespoons, drops, etc. The imperial measures are based on British weights and liquid measures. Note that American measures are different. For example, a standard American cup will hold 4 oz of sifted flour as compared with a standard British cup of 5 oz. Similarly, there are roughly 3 British tablespoons of sifted flour to the ounce as compared with 4 American tablespoons to the ounce.

The table below summarises the approximate capacities of the informal measures around the kitchen.

1 teaspoon	= 5 ml		
3 teaspoons	= 1 tablespoon (tbsp)		
1 tablespoonful	= 15 ml		
1 teacupful	= $\frac{1}{3}$ pint	= 7 fluid ounces	= 190 ml
1 breakfastcupful	= $\frac{1}{2}$ pint	= 10 fluid ounces	= 280 ml

There is no exact whole number conversion between metric and imperial measures, so whatever value you choose will depend on how accurate you need to be. In cooking, the needs for accuracy are usually not great, and indeed you wouldn't be able to weigh out ingredients to great accuracy anyway. The table below gives accurate and approximate conversions between imperial and metric measures.

Weight

To convert	Multiply by	
	Accurate figure	Cooking approximation
Ounces to grams	28.350	25
Pounds to grams	453.592	450
Pounds to kilograms	0.4536	0.45
Grams to ounces	0.0353	0.035
Grams to pounds	0.0022	0.022
Kilograms to pounds	2.2046	2.2

Liquid measures

To convert	Multiply by	
	Accurate figure	Cooking approximation
Pints to millilitres (ml)	568	550
Pints to litres (l)	0.568	0.55
Fluid ounces to ml	28.4	25
Fluid ounces to litres	0.0284	0.025
Millilitres to pints	0.00176	0.0017
Litres to pints	1.760	1.75
Millilitres to fluid ounces	0.0352	0.035
Litres to fluid ounces	35.21	35

How accurate do I need to be in my cooking?

This is a difficult question to answer precisely. With some recipes, for example vegetable soup or a salad mix, it isn't critical if you don't use the exact proportions stated in the recipe book. But if you are making, say, a subtle sauce (creamy paprika dressing, for example) the flavour could be affected by even a small error in one of the ingredients.

Have a look now at the basic ingredients for Bread and butter pudding, as given in my recipe book, and see if you can spot some sources of error in the measurement of these ingredients.

Bread and butter pudding
Thin slices of wholemeal bread 4 (about 4 oz/100 g)
Butter or margarine 1 oz (25 g)
Raw brown sugar 1 tbsp (15 ml)
Mixed sultanas, raisins & currants 2 oz (50 g)
Fresh milk $\frac{3}{4}$ pt (426 ml)
Free-range eggs 2
Ground cinnamon $\frac{1}{4}$ tsp (1.25 ml)
Nutmeg $\frac{1}{4}$ tsp (1.25 ml)
Serves 4

Here are a few points to note.
a Certain tiny amounts, like the 1.25 ml of nutmeg and ground cinnamon, are too small to weigh on kitchen scales. So you really will need to resort to using the informal measure of a $\frac{1}{4}$ tsp. It is actually unclear what this looks like. Recipes sometimes talk about a 'flat' teaspoonful and a 'heaped teaspoonful', so an ordinary teaspoonful is somewhere between the two. Measuring out a quarter of one of those is no easy task. The truth is that this sort of measure is very

approximate indeed and cooks will put in a variable amount of cinnamon and nutmeg, depending on whether or not they are keen on these flavours in their bread and butter pudding.

b Butter and margarine are rarely weighed out. Apart from the fact that they are difficult to weigh out as they tend to smear the weighing pan, it isn't necessary to do so. The standard method is to take a fresh pack of butter or margarine, which weighs, say, 500 g, and mark it out into five equal sections, thus:

50 g				
25 g				100 g
25 g				

Each main section will therefore be 100 g. Then take half and half again of one 100 g strip and this is 25 g.

c The amount of egg in the pudding will depend on the size of eggs used and egg size is not specified in the recipe. There is a considerable variation in egg weight, from 'very large' (73 g + over) down to 'small' (53 g + under). Egg sizes are classified into four weight bands, as follows.

Size	Weight
Very large	73 g + over
Large	63 – 73 g
Medium	53 – 63 g
Small	53 g + under

If you assume that a given 'very large' egg weighs 75 g and a given 'small' egg weighs 50 g, there is 50% more in the 'very large' egg than the 'small' egg. Looking at it another way, three 'small' eggs weigh roughly the same as two 'very large' eggs.

d Finally, have a look at the imperial and metric measures in this recipe (and others in your own recipe book). As was explained earlier, the conversions are only approximate. But just how approximate are they? The answer is that some are more aproximate than others. It is possible to calculate the percentage error of the conversions and this is shown in the table below.

Imperial	Stated metric	Actual metric	Error	Error (per cent)
4 oz	100 g	113.4	13.4	$\frac{13.4}{113.4} \times 100 = 12$
1 oz	25 g	28.35	3.35	12
$\frac{3}{4}$ pt	426 ml	426	0	0
$\frac{1}{2}$ pt	300 ml	284	16	6
1 lb	450 g	453.592	3.592	0.8

As is clear from the table, some conversions contain a substantial error (for example, the standard conversion from ounces to grams is 12% out) while others, like the $\frac{3}{4}$ pt of milk contain no evidence of error. Of course, whether you are able, accurately, to measure out exactly 426 ml of milk in your measuring jug is another question!

How do I scale up a recipe?

The recipe for bread and butter pudding given earlier serves four people, but I often cook for seven. This requires having to multiply each amount by the fraction $\frac{7}{4}$. The easiest way to do this is to set the calculator constant to × 1.75. (Using the calculator constant was explained in Chapter 03.) The results can then be rounded *sensibly*.

Original recipe	*Scaled up*	*Rounded*
Thin slices of wholemeal bread 4 (about 4 oz/100 g)	100 × 1.75 = 175 g	175 g
Butter or margarine 1 oz (25 g)	25 × 1.75 = 43.75 g	50 g
Raw brown sugar 1 tbsp (15 ml)	15 × 1.75 = 26.25 ml	25 ml
Mixed sultanas, raisins and currants 2 oz (50 g)	50 × 1.75 = 87.5 g	100 g
Fresh milk $\frac{3}{4}$ pt (426 ml)	426 × 1.75 = 745.5 ml	750 ml
Free-range eggs 2	2 × 1.75 = 3.5	4 small/ 3 large
Ground cinnamon $\frac{1}{4}$ tsp (1.25 ml)	$\frac{1}{4} \times \frac{7}{4} = \frac{7}{16}$ tsp	$\frac{1}{2}$ tsp
Nutmeg $\frac{1}{4}$ tsp (1.25 ml)	$\frac{1}{4} \times \frac{7}{4} = \frac{7}{16}$ tsp	$\frac{1}{2}$ tsp

Note: Sensible rounding of the larger metric numbers means rounding to the nearest 25 g or 25 ml. With ingredients like eggs, you can't easily add fractions of an egg, but you may have some flexibility over the size of eggs you use – for example, in this case, 3.5 eggs may approximate to either 4 small eggs or 3 large ones. But if you are like me, you just have one size of egg in your fridge and so you are stuck with what you've got!

Appendix H: Buying a TV set

Something like 96 per cent of households in the UK has (at least one) colour television set. Each of these households has therefore taken a decision about whether to buy or rent. If they chose to buy, they had a further choice as to whether to pay it all off straight away or to put down a deposit followed by regular instalments. The instalment method is also known as 'buying on credit' or HP (hire purchase). This method of payment is a bit like taking out a loan and you should expect to be charged more for paying in this way than by buying your TV outright.

This example focuses on how much you are likely to pay for your TV set if you decide to 'buy on credit'.

0% finance

Some shops offer a deal whereby you can buy on credit but the amount you pay overall is the same as if you bought the item outright. This will be advertised as 0% finance or 0% interest. For example:

SONY 21" NICAM Stereo TV with Fastext

- ■ 51cm visible screen size.
- ■ Superb NICAM stereo sound.
- ■ Fastext for easy access to all Teletext services.

18 months 0% interest

Price £349.99

20% Deposit & 18 direct debit monthly payments of £15.56.

Before checking the interest payments, let's just consider how the 'size' of this television set has been described in the advertisement. It is given separately both in imperial units (21 inches) and metric units (51 cm). All such measurements refer to the length of the diagonal of the screen, measured from corner to corner. Now, let's confirm that 21 inches and 51 cm really are the same length. To convert from inches to centimetres, we multiply by 2.54, thus:

$$21 \times 2.54 = 53.34.$$

Hmm. This result of 53.34 cm doesn't match up very well with the 51 cm figure I was expecting, so I'm not quite sure which, if either, of these figures to believe.

Something else worth checking here is the claim that this method of payment by monthly instalments really does represent 0% interest.

The figures can be checked as follows:

First, for convenience, let's round the price of the TV set up to £350.

Deposit =	20% of £350 = 0.2 × 350	= £70
18 monthly payments of £15.56 =	18 × 15.56	= £280.08
	TOTAL	= £350.08

OK, so you pay an extra 9 pence (£350.08, as compared with £349.99), but basically the total amount paid out by the instalment method is the same as the cash price. This confirms the claim that this method of payment does represent 0% interest.

By the way, don't assume that the 0% interest deal is always the best. Stores offering such deals may actually have higher prices for similar products than their rivals who may be offering a higher interest rate. In other words, the cost of the loan may be included in the price.

APR

0% interest is good when you can get it, but usually there is some interest charge when paying on credit. It is useful to know exactly how much you are being charged, and to be able to compare the 'real' interest rate between different shops. Dealers charge a variety of different interest rates, subject to the size of the deposit and the length of the repayment periods. As a result, it can be difficult to compare the actual interest being applied from one dealer to another. In recent years, this problem has been solved by the fact that all retailers are legally required to publish the effective interest rate of each deal on offer, using a measure called the 'annual percentage rate' or APR. The APR is the percentage cost of the loan, calculated over a year. It is quite difficult to calculate as the buyer pays a bit back at a time. The main thing to remember about APR is that a higher rate means that you pay more. For example, an APR of 32% means that you pay out more than with an APR of 27%. In general, all other things being equal (such as price, quality, after-sales service, insurance, and so on) look for the deal offering the lowest APR.

Appendix I: Will it fit?

Purchases of large household items, like a sofa, cabinet, or kitchen unit are often closely linked to the questions, 'Where will I put it?', and 'Will it fit?'. Ideally, these questions are sorted out *before* you have parted with your money, and not after!

Moving house is another sitution where questions of arranging the furniture have to be made, made sensibly, and, ideally, made in advance of removal day.

A useful strategy for helping you to decide where items of furniture should go is to produce a scale drawing of the various rooms and to make cardboard cut-out models of the sofa, table, TV, shelving unit, etc. This enables you to try things out without any of the sweat of trying out the objects themselves *in situ*.

The task of producing scale models and a scale drawing is actually quite straightforward and fun to do.

You will need the following resources:
- several sheets of squared paper or graph paper
- scissors
- tape measure
- ruler, pencil and access to the back-of-an-envelope!

Then follow the steps below.

1 Draw a rough 'back-of-an-envelope' sketch of the room, marking on the significant features – windows, doors, chimney breast, etc., and make a note of 'fixed points' like electric sockets, TV aerial, etc.

2 Using a tape measure, measure all the lengths of the room or rooms that you think you will need for your scale drawing. You are recommended to use metric units (metres and centimetres) as these are easier to deal with on the drawing.
3 Get some squared paper or graph paper and decide on a suitable scale. Ideally the final drawing should take up most of this sheet of paper (if the drawing is too small, you won't have much confidence in decisions where the fit is rather tight). Make the scale drawing on the squared or graph paper.
4 Measure the length and the width of each major item of furniture that might go in the room. Make a 2-D drawing, to scale, of each item on another sheet of squared or graph paper. Write the name of each item on its appropriate scale drawing, and then cut the models out.

You are now ready to 'play'!

Here is how I went about it for one room in my 'des. res.', the bedroom.

1 & 2 I made a sketch of my bedroom, measured the dimensions and marked them on, as shown below.

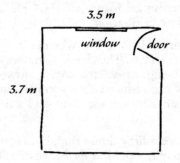

3 My squared paper is marked out in half centimetre squares. It is roughly 60 squares long and 40 squares wide. Since the bedroom is 3.7 m long and 3.5 m wide, I need to make a sensible judgement about the scale. (This is the only slightly tricky part of the job.) I decided to let 1 m = 10 squares. This meant that the room would be contained in a drawing of 37 squares by 35 squares.

4 The bedroom furniture was duly measured and again the same scale was applied. For example, the bed is 1.95 m long by 1.60 m wide. Using the scale of 1 m = 10 squares, this results in a cutout rectangle of 19.5 × 16 squares. The other items of furniture were cut out in the same way. *Note*: Care

needs to be taken with cupboards and cabinets in order that they are placed so that the doors are able to swing open. Similarly, it is helpful to mark the way the bedroom door opens, again to ensure that it is not obstructed.

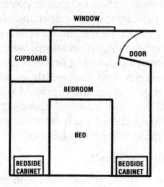

Appendix J: Measures of alcohol

Ethyl alcohol, chemical formula C_2H_5OH; 'an essence or spirit obtained by distillation'.

Not everyone drinks alcohol, but whether you do or not, you will be aware of its effects. The classic symptoms of the drug are a feeling of well-being, associated with a slowing down of the thought processes and reduced ability to react quickly. Taken in excess, alcohol can damage the liver and cause problems of overweight.

So far so bad! There is little doubt that, like cigarettes, if alcohol were invented today it would never be legalized!

If you or someone close to you does drink alcohol, it is sensible to know something about the alcoholic content of drinks and what sort of sensible limits are recommended by doctors.

Alcoholic content of drinks

Confusingly, there are two main ways of measuring how much alcohol there is in drink:

1 The old-fashioned measure of alcoholic strength is the degrees proof (e.g. 75° proof). This is measured in the range between a minimum of 0 and a maximum of 175. So water is 0° proof and neat alcohol would be measured at 175° proof.

This measure used to be applied to spirits and other drinks with a high alcohol content, but it is less commonly used these days.

2 Increasingly, bottles and cans of alcoholic drink are marked in terms of the percentage of alcohol in the drink (e.g. 8%). Calculated in this way, neat alcohol would be measured at 100 per cent. This measure has traditionally been applied to beers, cider, lager and other drinks with a relatively low alcohol content. However, most supermarkets and large retailers now use this method for spirits also.

The diagram below shows how to convert between these two measures.

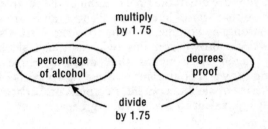

You might like to try the following exercise of converting between the two types of measure. And when you have completed the table, see if you can make some general comparisons between the strengths of different alcoholic drinks.

Drink	Degrees proof	Approximate % alcohol
White or red wine		14
Black Bush Irish Whiskey		40
Beauregard Napoleon Brandy		37.5
Safeway Vintage Port		20
Carlsberg Special (extra strength)	9.2	
Newcastle Brown Ale (strong ale)	7.5	
Woodpecker Cider (average cider)	5.5	
Tuborg Lager (ordinary lager)	3.7	

Solution

Drink	Degrees proof	Approximate % alcohol
White or red wine	24.5	14
Black Bush Irish Whiskey	70	40
Beauregard Napoleon Brandy	65.6	37.5
Safeway Vintage Port	35	20
Carlsberg Special (extra strength)	9.2	5.3
Newcastle Brown Ale (strong ale)	7.5	4.3
Woodpecker Cider (average cider)	5.5	3.1
Tuborg Lager (ordinary lager)	3.7	2.1

There are a few interesting points to emerge from the table. Running your eye down the final column of the completed table, you can see that spirits like whiskey and brandy are nearly 20 times as strong, by volume, as ordinary lager. Also, a strong ale like Newcastle Brown contains twice as much alcohol as an ordinary lager. This means that drinking three pints of Newcastle Brown is equivalent to drinking six pints of Tuborg. Also, in terms of alcohol content, two pints of Carlsberg Special is roughly equivalent to five of Tuborg Lager.

How do drinking habits vary around the UK?

The recommended maximum sensible amounts of alcohol are 21 units per week for men and 14 units per week for women. One unit is the equivalent of half a pint of ordinary strength beer or lager, a single measure of spirits, or a glass of wine. With this information and the data given in the table above, you should be able to work out how much of their favourite tipple various individuals should limit themselves to. For example, how many should the following individuals set as their upper weekly limits?

1 Hamish. He drinks Carlsberg Special extra strength lager.
2 Marti. She drinks average strength cider.

Solution
1 Hamish's rations
   Extra strength lager is roughly $2\frac{1}{2}$ times as strong as ordinary lager or beer (you can work this out by dividing the percentage alcohol content of extra strength lager, 5.3, by the percentage alcohol content of ordinary lager, 2.1. So, while Hamish would be able to drink 10.5 pints (21 units or half pints) of ordinary lager, the equivalent number of pints of

extra strength lager is calculated as $\frac{10.5}{2.5}$ = 4.2 pints. In other words, he should limit himself to a couple of pints per night, two nights a week.

2 **Marti's rations**
As a woman, Marti is allowed 7 pints (i.e. 14 units) of ordinary lager or beer. However, cider is stronger than lager. To calculate how much stronger, we must do the following calculation:

$\frac{3.1}{2.1}$ = 1.5 (approximately)

The equivalent number of pints of cider is calculated as $\frac{7}{1.5}$ = 4.7 pints.

In other words, she should limit herself to, say three half pints per night, three nights a week.

You may be wondering to what extent people do restrict their drinking within these upper limits. In general, women are more responsible in their drinking than men. There are also quite wide variations by region around the UK, as the graph below shows.

Typical consumption of alcohol above sensible limits[1]: by sex and region, GB

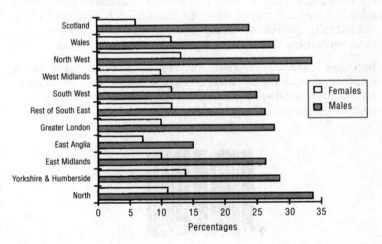

Source: *Social Trends, 24*

[1]Persons aged 16 and over consuming 22 units or more for males, and 15 or more units for females, per week.

One of the reasons that women are more affected by drink than men is that the water content of their bodies is differently constituted. In women, between 45 and 55 per cent of their body weight is made up of water. For men, the corresponding percentage is between 55 and 65. Since alcohol is distributed through the body fluids, so in men the alcohol is more 'diluted' than it is in women.

A second reason is that a woman's liver is more likely to suffer damage through alcohol poisoning than a man's.

Appendix K: Understanding barcodes

Up until the 1970s, supermarket goods were individually priced. Looking back, it is clear that this system had a number of drawbacks. Firstly, if the store decided to increase the price of, say, their baked beans, someone was required to collect all existing tins on the shelves, remove the existing labels and reprice each tin individually. Secondly, the system was open to abuse from dishonest customers who could switch price labels, substituting a cheaper one for a more expensive one before taking it through the check-out. Thirdly, each item had to be individually entered manually into the till at the check-out – a time-consuming task that was prone to error and abuse.

Barcodes changed all that. Instead of a price label being attached to each tin of beans or bag of muesli, etc., most items are manufactured with a barcode included on the packaging. A barcode looks something like this.

Barcoded items are scanned electronically, a process which is almost error-free. So any information that is encoded in the barcode is transferred via the scanner into the computerised till.

The beauty of the system is that the computer is able to log in much more information that simply the item's price. For example, each tin of beans that passes across the scanner is sold to a customer. This fact is automatically logged into the computer so that the store has a running count of their stock at any given time. At the end of each day they can then re-order new stocks of beans with some degree of precision. Precision in re-ordering is an important component in running a successful and competitive supermarket. Ordering insufficient tins of beans means the store may run out next day. Ordering too many results in a warehousing problem in storing the crates of surplus beans.

Sales over many months and years provide a valuable database of information from which the store can predict seasonal patterns and so fine-tune their re-orders. Also, price changes of an entire line can be entered as a single instruction on the store's computer without staff having to reprice existing stock, item by item, on the shelves.

As you can see if you examine any barcode, the bars are written alongside a row of numbers. When the electronic scanner 'reads' the bars, the information being inputted is actually these numbers in coded form. There are different barcode systems; some have just 8 digits while others have 13. Here is a 13-digit barcode for a 450g tin of Heinz baked beans.

5 000157 004185

These thirteen digits have been grouped rather oddly with the first digit, the 5, on its own and the remaining twelve digits split into two groups of six. This is how the human eye sees the number, but the computer scanner groups them differently. In terms of what information the computer needs, the thirteen digits split into four basic components, which are explained below.

50   The first two digits indicate the country of origin; in this case 50 means the UK.

00157   The next five digits refer to the manufacturer; the number allocated to all Heinz products is 00157.

00418   The next five digits indicate the particular product. So, Heinz have allocated these five digits to refer to a 450g tin of Heinz baked beans.

5   The final digit is known as a 'checksum'. It's purpose is to confirm that the digits recorded so far by the scanner are consistent and therefore likely to be correct. If the

scanner should read the first twelve digits given above, followed by any digit other than 5, the computer will record error (usually sounding a bleep) and the operator will know to rescan that item. The checksum is based on a formula applied to the previous twelve digits which should produce a single digit – in this case the number 5. The formula used for calculating this checksum is explained below.

The checksum

There are different ways of calculating checksums. This one is calculated by completing the following stages.

Stage	Example
1 Number the first twelve digits from 1 to 12.	1 2 3 4 5 6 7 8 9 10 11 12 5 0 0 0 1 5 7 0 0 4 1 8
2 Add together all the odd-numbered digits.	$5 + 0 + 1 + 7 + 0 + 1 = 14$
3 Add together all the even-numbered digits.	$0 + 0 + 5 + 0 + 4 + 8 = 17$
4 Now add the result of Stage 2 – the 14 – to three times the results of Stage 3 – the 17.	$14 + 3 \times 17 = 65$
5 Subtract the result of Stage 4 from the next bigger multiple of ten*, which in this case is 70.	$70 - 65 = 5$ This gives the checksum, 5, which becomes the thirteenth digit.

Note: If Stage 4 had produced the result 82, you would subtract this number from 90; a result of 56 would have to be subtracted from 60, and so on. In the case where the formula produces a result ending in zero (say 60) then subtract it from itself, producing a checksum of 0 (60 – 60).

The reason for setting up Stage 5 of the calculation in this form is to ensure a single-digit answer for the checksum.

You might like to explore this now for yourself. If you can't immediately lay your hands on any examples of 13-digit barcodes, then look at page 202 where one is reproduced. And here are two more to investigate.

Source	Barcode
Guardian newspaper 26 June 2001	9 770261 307729
Guardian newspaper 29 June 2001	9 770261 307750

- Which digits indicate that these newspapers were sold three days apart?
- Confirm that each checksum is correct.
- Suppose that a scanner misreads the first twelve digits of a barcode. How likely is it that the checksum would turn out to be correct for the incorrectly scanned number by chance alone?

Solutions

The final digits of these barcodes are the checksums. The only other differences in the codes are in the twelfth digits. Notice that digits 8 to 12 inclusive of the barcode for *Guardian* of 29 June are 30775, while those for 26 June are 30772, a difference of 3.

The checksums are calculated as follows.

26 June
$(9 + 7 + 2 + 1 + 0 + 7) + 3 \times (7 + 0 + 6 + 3 + 7 + 2) = 26 + 75 = 101$. $110 - 101 = 9$ Check!

29 June
$(9 + 7 + 2 + 1 + 0 + 7) + 3 \times (7 + 0 + 6 + 3 + 7 + 5) = 26 + 84 = 110$. $110 - 110 = 0$ Check!

Finally, there is a one in ten chance of the checksum being accepted even if the previous digits were scanned incorrectly. This is because there are ten possible digits available and there is therefore a one in ten chance that the checksum digit scanned happened to match the other twelve by chance alone.

Appendix L: Junk mail and free offers

They say there is no such thing as a free lunch and the same principle probably applies to free offers, especially when they take the form of unsolicited junk mail. Many of the offers that arrive on your mat are packaged in the form of a game of chance which you are invited to play. If you should be successful, and it's a fair bet that you will be, you qualify for their amazing free offer by being one of the very few lucky winners. You only need to complete the form and send away for big prizes. Further reading of the small print will no doubt reveal that things are not quite that simple.

A good example of this came through my door recently.

'Play this game and see how many mystery gifts you can claim!', the card read. The game card took the form of a 3×3 grid. Each cell in the grid was covered by a tear-off tab. The punter is asked to pull 3 tabs only. This revealed a number in each cell. If the numbers revealed added up to

6 .... claim 1 gift
7 .... claim 2 gifts
8 .... claim 3 gifts

My score came to 10, so I was clearly in luck. However, I couldn't resist tearing off the other six tabs (thereby making my game card void, but then that's life, eh!). The nine un-tabbed cells produced the following contents.

2	5	1
4	5	5
5	3	5

What this revealed was that the worst possible score I could get was $1 + 2 + 3 = 6$. In other words, I couldn't fail to win at least one prize. The second worst score I could get was $1 + 2 + 4 = 7$, which guaranteed two prizes. Any other combination guaranteed the maximum of three prizes. But just how many combinations are there altogether?

To calculate the number of possible selections, we go through each choice of cell in turn. There are 9 possible ways of choosing the first cell, 8 possible ways of choosing the second and 7 possible ways of choosing the third. Thus, the number of possible selections that I could have chosen is $9 \times 8 \times 7 = 504$. However, we need to be careful here, because each of these selections is the same as a number of other selections taken in a different order. In fact for any selection of three things, there are six possible orderings.

If you aren't convinced of this, consider a selection of the letters ABC. The six different ways of ordering these are as follows.

ABC
ACB
BAC
BCA
CAB
CBA

So, we conclude that the figure of 504 is actually six times too large if you wish to count only the number of possible combinations, without taking account of order. This gives a final figure of $\frac{504}{6}$ = 84 possible combinations from the game card.

So, it seems that, assuming the punters really do choose their tabs randomly, out of every 84 tries, the organizers could expect 1 person to win one prize, 1 to win two prizes and 82 to win all three. Put another way, something like $\frac{82}{84}$ = 0.976, or 97.6% of punters will get the thrill of hitting the jackpot on this game.

Hmm! Maybe I wasn't quite as lucky as I thought!

Appendix M: Winning on the National Lottery

In case you have never bought a lottery ticket, here is how it works.

Numbers from 1 to 49 inclusive are printed out on a 'pay slip'.

You choose six numbers between 1 and 49. If at least three of the numbers you choose match any of the six main numbers drawn, you are a winner.

Prizes vary depending on how many numbers you can match. At the time of writing, the prizes are:

Winning selections	Expected prize
■ Jackpot. Match 6 main numbers	2 million
■ Match 5 main numbers plus the bonus number	£100 000
■ Match 5 main numbers	£1 500
■ Match 4 main numbers	£65
■ Match 3 main numbers	£10

Most countries run national lotteries. They provide a lot of fun and fantasy for the punters, and, of course, are nice little earners for the government.

Lottery fever hit the UK in November 1994, with a much hyped launch on TV, radio and the press. A lot of advice and information was offered to the great British public, most of which was total nonsense.

On the Jonathan Ross television show, a clairvoyant, named The Voxx, came up with a photo-fit of the winner:

> '...in her forties with strawberry blond hair, possibly dyed, of Irish or Scottish background, has travelled extensively and has had a tough time in love but there is someone in her life at the moment. She also has a son.'

Well, that should narrow it down to a few hundred thousand people!

The Sun newspaper helpfully printed a giant dot, charged with lucky psychic energy. Readers were invited to touch the lucky spot, close their eyes and the numbers would just come to them by the sheer power of the 'Lottery spottery'. *The Sun* provided some evidence from America (where lucky spots were first devised) for this claim. Apparently, 'thousands of people said they only won because of their special power'. Wow!

Other papers offered yet more advice. For example, avoid so-called 'lucky' numbers like 7, 11 and 13, and avoid choosing numbers relating to birthdays or anniversaries. Can you think of any rational explanation for this?

So, who is to be believed and is there a 'best strategy' for playing the lottery? Fortunately in answering these questions, mathematics can provide insights to some of the parts that even lottery spottery cannot reach!

Let's try to sort out some of the fact from the fiction. Two key ideas will be explained below. The first explores the chance of winning – what sort of odds you are really up against. The second is to do with the way the numbers are chosen, both by the lottery 'random number generator' (the name for the machine that spits out the winning numbers) and how the numbers are selected by the paying, playing punters.

What are my chances?

Roughly half of the money paid into the lottery is given back in prizes. So, taking a very long term view, if you bought, say £1000 worth of lottery tickets over your lifetime, you could expect, on average, to lose about £500. The reality is that you will almost certainly not win one of the monster prizes, but then again, you just might. According to the promoters, the odds against winning the jackpot of, say, £2 million (although this figure depends on how many people play) are about 14 million to one. By the way, if you are interested in how this figure of 14 million is calculated, it is explained on page 211. So for this prize, you would expect, on average, to have to lay out £14 million to win back £2 million. At the other end of the winnings scale, there is one chance in 57 of winning a £10 guaranteed prize – i.e. you would expect, on average to lay out £57 to win back £10.

In terms of return on your investment, this is pretty thin gruel, whichever way you serve it up. But then again, what keeps most of us losing money on such foolishness is that we might just win that big one…!

How are the numbers chosen?

As with Bingo, a key principle of the lottery is equal likelihood – i.e. the device for choosing the numbers is designed so that each number has an equal chance of coming up. The only thing that would prevent that from happening is if the random number generator was programmed to generate numbers in a different way. But then that would be cheating.

Numbers which have an equal chance of coming up are known as *random numbers* (for example, tossing dice, coins, and so on). So, since lottery numbers are chosen at random, no number or combination of numbers is more or less likely to come up than any other. For example, the selection 1, 2, 3, 4, 5, 6 is just as likely (well, just as *unlikely* would be more appropriate) as a mixed bag of numbers like 32, 6, 18, 41, 9, 15.

Clearly, then, you have no control over whether or not your numbers win. But what about sharing your winnings with others? Here you *can* exercise your skill and judgement by anticipating what numbers other punters are likely to choose.

Basically, if you come up with a winning combination, you will share it with fewer people if you pick numbers that others are less likely to pick. So this is where a bit of mind reading comes in. A simple example may make this clearer.

A simplified lottery example

Twenty punters, paying £1 each to play, choose a number in the range 1 to 6. A die is tossed and the £10 winnings[1] are shared among those who chose the winning number.

The punters' choices are shown below.

Selection	1	2	3	4	5	6
No. of people	2	5	4	5	3	1

In other words, two punters chose '1', five chose '2', four chose '3', and so on. Now, notice that if the die shows up '2' or '4', the winnings have to be shared among five winners, so each of these winning punters gets $\frac{£10}{5}$ = £2. But it is just as likely that the winning number turns out to be '6', in which case the lucky winner scoops the lot. In fact, '6' would be a good choice here, because (due to negative experiences playing board games) people tend to avoid this number in the mistaken belief that it is less likely to come up than any other.

So, a good overall strategy is to avoid numbers that other people are likely to choose and to go for numbers that they are unlikely to choose.

As far as the National Lottery is concerned, that means:

Avoid	people's 'lucky' numbers, like 3, 7, 11.
Avoid	numbers linked to birthdays or anniversaries. In other words avoid all numbers of 31 or below (days in a month) and particularly avoid numbers of 12 or below (months in a year).
Go for	numbers above 31.
Go for	numbers in a sequence (many people mistakenly think that these are less likely than numbers which 'look random').

---

[1]'Hey, where did the other £10 go?'
'To the lottery organisers, of course. There are so many expenses; they have huge advertising and administrative costs. Then they have to buy the die, and that doesn't come cheap. And, of course, they need to train their staff to toss it and see fair play all round.'

Finally, note that these strategies are only successful if we have successfully predicted how the other punters will choose their numbers (i.e. that they will tend to go for numbers less than 31, avoid numbers in sequence, and so on). In fact all the indications are that this is indeed what people do. Evidence for this emerged from the very first UK national lottery in November 1994 which produced the following winning numbers

3, 5, 14, 22, 30 and 44 (10 was the bonus number)

To the great disappointment of the organizers, Camelot, the jackpot had to be scaled down from an estimated £7m to £5.8 as the number of small-scale winners became known. Camelot staff had not expected there to be as many as the one million players who would pick up the guaranteed £10 prize pay-out for three correct numbers. The reason, it seems, was that most of the players did pick numbers relating to birthdays and as can be seen, five of the six winning numbers were below 31.

So, best strategy at the time of writing, is to choose numbers above 31 and opt for strings, rather than avoid them as the naive players will. Of course, at a certain point, when enough people have read this book, it may be the case that the informed punters will outnumber the naive ones. When that happy state arrives, you may feel it appropriate to change your strategy accordingly. Me? Well, if this knowledge is brought about by my book selling more than 5 million copies, I certainly won't need to waste my time or money on a dumb lottery!

How is the figure of 14 million to one calculated?

It is correctly claimed by the lottery organizers that if you choose six numbers at random from a list of 49, the odds are about 14 million to one against your selecting the six winning numbers. The calculation is explained below based on an initial slight incorrect assumption, but I will sort that out at the end!

Start by choosing the first number. There are 49 to choose from so there are clearly 49 choices. Choosing the second number means choosing from the 48 remaining numbers. So, *for each* of the 49 first choices there are 48 second choices. In other words, there are 49×48 ways of choosing the first two numbers. Similarly, there are $49 \times 48 \times 47$ ways of choosing the first three numbers, and so on. Following the same line of argument, it would seem that the number of ways of choosing the first six numbers correctly are:

$$49 \times 48 \times 47 \times 46 \times 45 \times 44$$

Unfortunately, most basic calculators are unable to perform this calculation because it produces a result too big for the calculator to display. So will you take my word for it that this is the correct answer? Well, you shouldn't! As I suggested earlier, there is an error in the reasoning here, which I will now correct.

The error is that I have included each combination of numbers many times, as a result of which this answer is too large. To convince you of this, imagine that the first set of six numbers you chose was:

1, 2, 3, 4, 5 and 6

This same combination of numbers could also crop up as 1, 3, 2, 4, 5, 6 or 6, 5, 4, 3, 2, 1 or any ordering you can think of. But just how many orderings are there? This question was explored in Appendix L: Junk mail and free offers. There we ordered three things and found that there were six possible orderings. The number of ways of ordering six things is more complicated and can be calculated as follows.

There are six ways of ordering the first number, five ways of ordering the second, four ways of ordering the third, and so on.

So, the number of ways of ordering six things = $6 \times 5 \times 4 \times 3 \times 2 \times 1 = 720$.

And now, back to the plot. This last discussion suggests that *each combination* of numbers contained in the calculation

$$49 \times 48 \times 47 \times 46 \times 45 \times 44$$

is actually included 720 times.

So, the number of separate combinations = $\frac{49 \times 48 \times 47 \times 46 \times 45 \times 44}{720}$

Now we can punch this out on the calculator and hope we get the answer 14 million. One snag is that, if you calculate the top line first before dividing by 720, you will almost certainly cause the calculator to overflow. A sneaky way out of this is to divide by the 720 sooner rather than later in the calculation. For example, I pressed the following:

49 ÷ 720 × 48 × 47 × 46 × 45 × 44 =

This produces the answer 13 983 816, which isn't all that far away from the result we were hoping for of 14 million.

Appendix N: Safe travel

Is it safer to travel by road or by rail?

Most people say that rail travel is safer, but is this really true?

A useful start in answering this question is to look at the number of deaths in a year due to railway and road accidents. Typical annual figures for Great Britain, supplied by the Office of Population Censuses and Surveys (OPCS) are as follows:

Deaths due to railway accidents	70
Deaths due to road accidents	4628

So, clearly, many more people die on the roads than by travelling on a train – in fact about seventy times as many, in the year in question. But does this mean that rail travel is seventy times safer than road travel? The answer is, not necessarily, because we may not be comparing like with like. An important complicating factor is that many more people travel many more kilometres by road than by rail. For example, private motor vehicles and taxis are used for around 90 per cent of distances travelled in the UK. To answer the question fairly, we need to take account of the average distances travelled by each mode of transport and use these to calculate the accident *rates*. These can then be compared directly. So here goes.

The fairest figure to use here is the number of passenger kilometres for both road and rail. As the name implies, one passenger kilometre is recorded when one passenger travels one kilometre. If five passengers each travel 10 kilometres, a total of 50 passenger kilometres will be recorded.

Typical annual rail passenger transport use	= 100 billion passenger km
Typical annual road passenger transport use	= 590 billion passenger km

To calculate the accident rate per billion passenger kilometres, we divide the number of deaths in a year by the number of billion passenger kilometres, thus:

Rail passenger death rate $= \frac{70}{100} = 0.7$
Road passenger death rate $= \frac{4628}{590} = 7.8$

Using this comparison, it seems that road travel is roughly ten times as dangerous as rail travel.

It is interesting to look at other forms of transport, based on this comparison of the number of deaths per billion passenger kilometres. Typical annual figures for Great Britain are as follows.

Mode	Rate per billion passenger kilometres
Air	0.1
Water	0.5
Rail	0.7
Bus or coach	0.4
Car	3.6
Van	2.2
Motorcycles	97.0
Pedal cyclists	43.4
Foot	53.4

Source: Department of Transport

So, according to this measure, air travel really is the safest form of transport and motorcycling is very much the most dangerous. However, it should be stressed that all measures have their drawbacks and this is just one possible measure. A weakness in the measure used here is that it favours modes of travel which are fast over the slower methods. Thus, air travel can allow you to cover, say, a hundred kilometres in just a few minutes, whereas you would take days on foot to cover this sort of distance. As a result, for a given number of passenger kilometres, the exposure to risk on foot is much greater merely due to the fact that there is a longer period of time during which an accident can happen. However, provided this aspect is borne in mind, using the number of deaths per billion passenger kilometres is probably the fairest comparison available.

Appendix O: World population

It used to be said that everyone in the world could just fit onto the Isle of Wight if they all squeezed up a bit. How could you check a claim like this?

Clearly, it is impossible to prove or disprove such a claim with absolute certainty – for one thing, the facts and figures needed are simply not available with perfect accuracy and there are too many practical difficulties (are we allowed to knock down all the buildings and trees, for example?). But there is some fun to be had in making sensible guesses and doing an 'order of magnitude' calculation.

First, let us establish what information is needed to make the calculation.

- What is the current population of the world?
  Clearly it is changing all the time, but we can look it up in a reference book. According to the UK government publication *'Social Trends'*, the estimate of the world's population in 2001 was about 6.2 billion.
- Next, what is the area of the Isle of Wight?
  According to the *Macmillan Encyclopedia*, this is 380 sq km (147 sq miles).
- Finally, we need to make a sensible guess as to how many people could be fitted into one square metre of space.

  Assuming they are all standing up (and breathing in) let's guess that ten people could be squeezed into such a space.

So, now for the calculation.

Total area, in square metres $= 380 \times 1000\,000$
Number of people who would fit into
  this area $= 380 \times 1000\,000 \times 10$
$= 3.8$ billion

Since this figure is less than the 6.2 billion who are estimated to populate the globe, the answer would appear to be that, even if all the trees, houses, cows and lamp-posts were to be removed, it simply couldn't be done.

As a footnote to this investigation, it is worth pointing out that this claim has been around for many decades, over which period the world's population has grown considerably. It is estimated that the population is increasing at a rate of roughly 18 per cent each decade. This means that estimates of future world populations, decade by decade, can be made by multiplying the present estimate by 1.18. (If you aren't sure where this figure of 1.18 came from, it is explained in Chapter 06.)

For example, to make an estimate of the population in 2011, multiply by 1.18, thus:

$$6.2 \text{ billion} \times 1.18 = 7.3 \text{ billion}$$

We can make backward projections in a similar way, but dividing by 1.18 instead of multiplying.

An estimate of the population in 1991, 1981 and so on can be found as follows:

  1991 estimate = 6.2 billion $\div$ 1.18 = 5.25 billion
  1981 estimate = 5.25 billion $\div$ 1.18 = 4.45 billion
  1971 estimate = 4.45 billion $\div$ 1.18 = 3.77 billion

So it would seem that, in 1971, the claim wasn't entirely preposterous. And there might even have been enough room for the cows!

Completed police file

We have reason to believe that the suspect entered the premises of *Devlin's* shop at *The Shires Park* in the town of *Doddington* on the afternoon of *24th Dec* in the year *2001*. At precisely *4.26 p.m.*, 2 items were purchased, namely a *bottle of dishwasher liquid* and a *bunch of tulips,* costing £3.29 and £2.95 respectively.

A £10 note was submitted to the cashier and £3.76 in change was received.

Websites and organizations

Name/resource

A+B Books, who write and publish mathematics books for use with a graphics calculator.
www.AplusB.co.uk

Association of Teachers of Mathematics (UK)
www.atm.org.uk/

Bournmouth University applets (computer animations) demonstrating mathematical principles
http://mathinsite.bmth.ac.uk/html/applets.html

Coventry University Mathematics Support Centre for Mathematics Education
www.mis.cov.ac.uk/maths_centre/

Mathematical Association (UK)
www.m-a.org.uk/

Mathpuzzle contains a range of puzzles and other maths resources
www.Mathpuzzle.com/

MathsNet contains a wide range of puzzles, books and software
http://www.mathsnet.net/

National Council of Teachers of Mathematics (USA)
www.nctm.org

Open University Mathematics courses
www.open.ac.uk/maths

Oundle School site, with many useful resources and links to other sites worldwide
http://www.argonet.co.uk/oundlesch/mlink.html

UK National Statistics on line
http://www.statistics.gov.uk/

University of Plymouth Mathematics support Materials
www.tech.plym.ac.uk/maths/resources/PDFLaTeX/mathaid.html

Reading list

Barrow, John D. (1993) *Pi in the Sky: Counting, Thinking and Being*, Penguin, London. An exploration of where maths comes from and how it is performed.

Eastaway, Rob & Wyndham, Jeremy (1998) *Why Do Buses Come in Threes?*, Robson Books, London. Practical uses for various mathematical topics, including probability, Venn Diagrams and prime numbers.

Eastaway, Rob & Wyndham, Jeremy (2002) *How long is a piece of string?*, Robson Books, London. Examples of mathematics in everyday life.

Flannery, Sarah (2000) *In Code: A Mathematical Journey*, Profile Books, London. A collection of problems with solutions and explanations, based on the author's experiences of growing up in a mathematical home.

Graham, Alan (2003) *Teach Yourself Statistics*, Hodder & Stoughton, London. A straightforward and accessible account of the big ideas of statistics with a minimum of hard mathematics.

Huntley, H.E. (1970) *The Divine Proportion, Study in Mathematical Beauty*, Dover, New York. Applications in art and nature of the 'Golden Ratio'.

Ifrah, Georges (1998) *The Universal History of Numbers*, The Harvill Press, London. A detailed book (translated from French) about the history of numbers and counting from pre-history to the age of the computer.

Paulos, John Allen (1990) *Innumeracy – Mathematical Illiteracy and its Consequences*, Penguin, London. Real-world examples of innumeracy, including stock scams, risk perception and election statistics.

Pólya, G. (1990) *How to Solve It*, Penguin, London. A classic text on mathematical problem solving that is well-known around the world.

Singh, Simon (2000) *The Code Book*, Fourth Estate, London. A history of codes and ciphers and their modern applications in electronic security.

Singh, Simon (1998) *Fermat's Last Theorem*, Fourth Estate, London. An account of Andrew Wiles' proof of Fermat's Last Theorem, but also outlining some problems that have interested mathematicians over many centuries.

Stewart, Ian (1996) *From Here to Infinity*, Oxford University Press, Oxford. An introduction to how mathematical ideas are developing today.

Stewart, Ian (1997) *Does God Play Dice?*, Penguin, London. An introduction to the theory and practice of chaos and fractals.

index

teach® yourself

the A-Z of teach yourself

Afrikaans
Access 2002
Accounting, Basic
Alexander Technique
Algebra
Arabic
Arabic Script, Beginner's
Aromatherapy
Astronomy
Bach Flower Remedies
Bengali
Better Chess
Better Handwriting
Biology
Body Language
Book Keeping
Book Keeping & Accounting
Brazilian Portuguese
Bridge
Buddhism
Buddhism, 101 Key Ideas
Bulgarian
Business Studies
Business Studies, 101 Key Ideas
C++
Calculus
Calligraphy
Cantonese
Card Games
Catalan
Chemistry, 101 Key Ideas
Chess
Chi Kung
Chinese
Chinese, Beginner's

Chinese Language, Life & Culture
Chinese Script, Beginner's
Christianity
Classical Music
Copywriting
Counselling
Creative Writing
Crime Fiction
Croatian
Crystal Healing
Czech
Danish
Desktop Publishing
Digital Photography
Digital Video & PC Editing
Drawing
Dream Interpretation
Dutch
Dutch, Beginner's
Dutch Dictionary
Dutch Grammar
Eastern Philosophy
ECDL
E-Commerce
Economics, 101 Key Ideas
Electronics
English, American (EFL)
English as a Foreign Language
English, Correct
English Grammar
English Grammar (EFL)
English, Instant, for French Speakers
English, Instant, for German Speakers
English, Instant, for Italian Speakers
English, Instant, for Spanish Speakers

English for International Business
English Language, Life & Culture
English Verbs
English Vocabulary
Ethics
Excel 2002
Feng Shui
Film Making
Film Studies
Finance for non-Financial Managers
Finnish
Flexible Working
Flower Arranging
French
French, Beginner's
French Grammar
French Grammar, Quick Fix
French, Instant
French, Improve your
French Language, Life & Culture
French Starter Kit
French Verbs
French Vocabulary
Gaelic
Gaelic Dictionary
Gardening
Genetics
Geology
German
German, Beginner's
German Grammar
German Grammar, Quick Fix
German, Instant
German, Improve your
German Language, Life & Culture
German Verbs
German Vocabulary
Go
Golf
Greek
Greek, Ancient
Greek, Beginner's
Greek, Instant
Greek, New Testament
Greek Script, Beginner's
Guitar
Gulf Arabic
Hand Reflexology
Hebrew, Biblical
Herbal Medicine
Hieroglyphics
Hindi
Hindi, Beginner's
Hindi Script, Beginner's

Hinduism
History, 101 Key Ideas
How to Win at Horse Racing
How to Win at Poker
HTML Publishing on the WWW
Human Anatomy & Physiology
Hungarian
Icelandic
Indian Head Massage
Indonesian
Information Technology, 101 Key Ideas
Internet, The
Irish
Islam
Italian
Italian, Beginner's
Italian Grammar
Italian Grammar, Quick Fix
Italian, Instant
Italian, Improve your
Italian Language, Life & Culture
Italian Verbs
Italian Vocabulary
Japanese
Japanese, Beginner's
Japanese, Instant
Japanese Language, Life & Culture
Japanese Script, Beginner's
Java
Jewellery Making
Judaism
Korean
Latin
Latin American Spanish
Latin, Beginner's
Latin Dictionary
Latin Grammar
Letter Writing Skills
Linguistics
Linguistics, 101 Key Ideas
Literature, 101 Key Ideas
Mahjong
Managing Stress
Marketing
Massage
Mathematics
Mathematics, Basic
Media Studies
Meditation
Mosaics
Music Theory
Needlecraft
Negotiating
Nepali

Norwegian
Origami
Panjabi
Persian, Modern
Philosophy
Philosophy of Mind
Philosophy of Religion
Philosophy of Science
Philosophy, 101 Key Ideas
Photography
Photoshop
Physics
Piano
Planets
Planning Your Wedding
Polish
Politics
Portuguese
Portuguese, Beginner's
Portuguese Grammar
Portuguese, Instant
Portuguese Language, Life & Culture
Postmodernism
Pottery
Powerpoint 2002
Presenting for Professionals
Project Management
Psychology
Psychology, 101 Key Ideas
Psychology, Applied
Quark Xpress
Quilting
Recruitment
Reflexology
Reiki
Relaxation
Retaining Staff
Romanian
Russian
Russian, Beginner's
Russian Grammar
Russian, Instant
Russian Language, Life & Culture
Russian Script, Beginner's
Sanskrit
Screenwriting
Serbian
Setting up a Small Business
Shorthand, Pitman 2000
Sikhism
Spanish
Spanish, Beginner's
Spanish Grammar
Spanish Grammar, Quick Fix

Spanish, Instant
Spanish, Improve your
Spanish Language, Life & Culture
Spanish Starter Kit
Spanish Verbs
Spanish Vocabulary
Speaking on Special Occasions
Speed Reading
Statistical Research
Statistics
Swahili
Swahili Dictionary
Swedish
Tagalog
Tai Chi
Tantric Sex
Teaching English as a Foreign Language
Teaching English One to One
Teams and Team-Working
Thai
Time Management
Tracing your Family History
Travel Writing
Trigonometry
Turkish
Turkish, Beginner's
Typing
Ukrainian
Urdu
Urdu Script, Beginner's
Vietnamese
Volcanoes
Watercolour Painting
Weight Control through Diet and
  Exercise
Welsh
Welsh Dictionary
Welsh Language, Life & Culture
Wills and Probate
Wine Tasting
Winning at Job Interviews
Word 2002
World Faiths
Writing a Novel
Writing for Children
Writing Poetry
Xhosa
Yoga
Zen
Zulu

available from bookshops and on-line retailers

teach
yourself

mathematics
trevor johnson & hugh neill

- Do you want a step-by-step introduction to essential mathematical concepts and techniques?
- Do you like clear explanations and examples to guide you?
- Do you want to test your understanding with exercises and answers throughout the book?

Mathematics is a comprehensive introduction to the key areas of the subject. It guides you to an understanding of each concept, and reinforces your knowledge with exercises and examples. The book includes basic arithmetical processes, algebra and geometry, fractions and decimals, and much more. Ideal for those who want to gain both knowledge and confidence.

Trevor Johnson is an Examiner for EdExcel GCSE. **Hugh Neill** is a former Examiner for A Level Maths and SCAA Maths Consultant.

| teach yourself | **calculus**
hugh neill |

- Are you looking for a comprehensive introduction to calculus?
- Do you need the essential mathematical background to understand and apply it?
- Are you a beginner looking to progress from basics to a high standard?

Calculus provides a carefully graded series of lessons which introduce the underlying concepts of differentiation and integration. Each chapter includes many clearly worked examples, diagrams and exercises with answers. It is suitable for those studying pure and applied mathematics, engineering and allied sciences

Hugh Neill is a mathematics author and a teacher, inspector and chief examiner in mathematics at various levels.

teach yourself

trigonometry
p. abbott & hugh neill

- Are you new to trigonometry?
- Do you need practice for an exam or course?
- Do you need to refresh your understanding?

Trigonometry offers a comprehensive introduction to trigonometry which progresses steadily to more advanced skills. Worked examples and carefully graded exercises are supported by extensive answers including full demonstrations of trigonometric proofs.

Hugh Neill is a mathematics author and a former teacher, inspector and chief examiner in mathematics at various levels.

teach yourself	**statistics** alan graham

- Do you want to know how to interpret figures?
- Do you want to put statistics into everyday contexts?
- Do you need to know the key ideas and principles of the subject?

Statistics assumes no previous knowledge or mathematical background and provides help with basic maths as well as focusing on the key ideas and techniques of statistics. Nearly all aspects of human lives can be subject to statistical analysis and using this book you can learn how to interpret and present figures, following examples from a wide variety of everyday situations.

Alan Graham works at the Open University's Centre for Mathematics Education, and has a particular interest in graphic calculators and statistics education.